LECTURES ON COMPLEX ANALYTIC VARIETIES: FINITE ANALYTIC MAPPINGS

BY

R. C. GUNNING

PRINCETON UNIVERSITY PRESS

AND

UNIVERSITY OF TOKYO PRESS

PRINCETON, NEW JERSEY

1974

Published by Princeton University Press, Princeton and London

L.C. Card: 74-2969

I.S.B.N.: 0-691-08150-6

Library of Congress Cataloging in Publication Data
will be found on the last printed page of this book

Published in Japan exclusively
by University of Tokyo Press
in other parts of the world by
Princeton University Press

Printed in the United States of America

PREFACE

These notes are intended as a sequel to "Lectures on Complex Analytic Varieties: The Local Parametrization Theorem" (Mathematical Notes, Princeton University Press, 1970), and as in the case of the preceding notes are derived from courses of lectures on complex analytic varieties that I have given at Princeton in the past few years. There are a considerable variety of topics which can be treated in courses of lectures on complex analytic varieties for students who have already had an introduction to that subject. The unifying theme of these notes is the study of local properties of finite analytic mappings between complex analytic varieties; these mappings are those in several dimensions which most closely resemble general complex analytic mappings in one complex dimension. The purpose of these notes though is rather to clarify some algebraic aspects of the local study of complex analytic varieties than merely to examine finite analytic mappings for their own sake. Some of the results covered may be new, and in places the organization of the material may be somewhat novel. In the course of the notes I have supplied references for some results taken from or inspired by recent sources, although no attempt has been made to provide complete references. Needless to say most of the material is part of the current folklore in several complex variables, and the purely algebraic results in the third section are quite standard and well known in the study of local rings.

I should like to express my thanks here to the students who have attended the various courses on which these notes are based, for all of their helpful comments and suggestions, and to Mary Ann Schwartz, for a beautiful typing job.

Princeton, New Jersey R. C. Gunning

January, 1974

CONTENTS

Page

§1. Finite analytic mappings 1

a. Analytic varieties: a review (1)
b. Local algebras and analytic mappings (6)
c. Finite analytic mappings (11)
d. Characteristic ideal of an analytic mapping (18)
e. Weakly holomorphic and meromorphic functions (28)

§2. Finite analytic mappings with given domain 38

a. Algebraic characterization of the mappings (38)
b. Normal varieties and local fields (48)
c. Examples: some one-dimensional varieties (56)
d. Examples: some two-dimensional varieties (71)

§3. Finite analytic mappings with given range 86

a. Algebraic characterization of the mappings (86)
b. Perfect varieties and removable singularity sets (93)
c. Syzygies and homological dimension (100)
d. Imperfect varieties and removable singularity sets (109)
e. Zero divisors and profundity (117)
f. Profundity and homological dimension for analytic varieties (127)

Appendix. Local cohomology groups of complements of complex analytic subvarieties 144

Index of symbols 160

Index 161

§1. Finite analytic mappings

(a) These notes are intended as a sequel to the lecture notes CAV I*, so it will be assumed from the outset that the reader is somewhat familiar with the contents of the earlier notes and the notation and terminology introduced in those notes will generally be used here without further reference. It will also be assumed that the reader has some background knowledge of the theory of functions of several complex variables and of the theory of sheaves, at least to the extent outlined at the beginning of the earlier notes. For clarity and emphasis however a brief introductory review of the definitions of germs of complex analytic subvarieties and varieties will be included here.

A *complex analytic subvariety* of an open subset $U \subseteq \mathbb{C}^n$ is a subset of U which in some open neighborhood of each point of U is the set of common zeros of a finite number of functions defined and holomorphic in that neighborhood. A *germ of a complex analytic subvariety* at a point $a \in \mathbb{C}^n$ is an equivalence class of pairs (V_α, U_α), where U_α is an open neighborhood of the point a in $\mathbb{C}^n$, V_α is a complex analytic subvariety of U_α, and two pairs (V_α, U_α) and (V_β, U_β) are equivalent if there is an open neighborhood U of the point a in $\mathbb{C}^n$ such that $U \subseteq U_\alpha \cap U_\beta$ and

---

* Lectures on Complex Analytic Varieties: the Local Parametrization Theorem. (Mathematical Notes, Princeton University Press, Princeton, N. J., 1970.)

$U \cap V_\alpha = U \cap V_\beta$. A complex analytic subvariety V of an open subset $U \subseteq \mathbb{C}^n$ determines a germ of a complex analytic subvariety at each point $a \in U$, and this germ will also be denoted by V; consideration of the germ merely amounts to consideration of the local properties of V near the point a. Two germs V_1, V_2 of complex analytic subvarieties at points a_1, a_2 in $\mathbb{C}^n$ are equivalent germs of complex analytic subvarieties of $\mathbb{C}^n$ if they can be represented by complex analytic subvarieties V_1, V_2 of open neighborhoods U_1, U_2 of the respective points a_1, a_2 for which there exists a complex analytic homeomorphism $\varphi: U_1 \to U_2$ such that $\varphi(V_1) = V_2$ and $\varphi(a_1) = a_2$. Consideration of these equivalence classes merely amounts to consideration of the properties of germs of complex analytic subvarieties of $\mathbb{C}^n$ which are independent of the choice of local coordinates in $\mathbb{C}^n$; for this purpose the germs of complex analytic subvarieties can all be taken to be at the origin in $\mathbb{C}^n$.

A continuous mapping $\varphi: V_1 \to V_2$ from a germ V_1 of a complex analytic subvariety at a point $a_1 \in \mathbb{C}^{n_1}$ to a germ V_2 of a complex analytic subvariety at a point $a_2 \in \mathbb{C}^{n_2}$ is the germ at the point a_1 of a continuous mapping φ from a subvariety representing V_1 into a subvariety representing V_2 such that $\varphi(a_1) = a_2$. A continuous mapping $\varphi: V_1 \to V_2$ is a complex analytic mapping if the germs V_1, V_2 can be represented by complex analytic subvarieties V_1, V_2 of open neighborhoods U_1, U_2 of the respective points $a_1 \in \mathbb{C}^{n_1}$, $a_2 \in \mathbb{C}^{n_2}$ for which there is a complex analytic mapping $\Phi: U_1 \to U_2$ such that $\Phi(V_1) \subseteq V_2$, $\Phi(a_1) = a_2$, and φ is the germ at the point a_1 of the restriction $\varphi = \Phi|V_1$. Two germs V_1, V_2 are <u>topologically equivalent</u> if there are continuous mappings $\varphi: V_1 \to V_2$ and $\psi: V_2 \to V_1$ such that the compositions $\psi\varphi: V_1 \to V_1$ and $\varphi\psi: V_2 \to V_2$ are the identity mappings; this is of course just the condition that the germs V_1, V_2 have topologically homeomorphic representative subvarieties in some open neighborhoods of the points a_1, a_2. Two germs V_1, V_2 are <u>equivalent germs of complex analytic varieties</u> if there are complex analytic mappings $\varphi: V_1 \to V_2$ and $\psi: V_2 \to V_1$ such that the compositions $\psi\varphi: V_1 \to V_1$ and $\varphi\psi: V_2 \to V_2$ are the identity mappings; and an equivalence class is a <u>germ of a complex analytic variety</u>. It is evident that this is a weaker equivalence relation than that of equivalence of germs of complex analytic subvarieties; thus there is a well defined germ of complex analytic variety underlying any germ of complex analytic subvariety, or indeed any equivalence class of germs of complex analytic subvarieties. The germ of complex analytic variety represented by a germ V of complex analytic subvariety will also be denoted by V. The distinguished point on a germ of complex analytic variety will be called the <u>base point</u> of the germ, and will be denoted by 0; for a germ of complex analytic variety can always be represented by a germ of complex analytic subvariety at the origin in some complex vector space. It is also evident that equivalent germs of complex analytic varieties are topologically equivalent; thus there is a well defined germ of a topological space underlying any germ of a complex analytic variety.

To any germ V of a complex analytic subvariety at a point $a \in \mathbb{C}^n$ there is associated the ideal $\text{id}\, V \subseteq {}_n\mathcal{O}_a$ consisting of those germs of holomorphic functions at the point a in $\mathbb{C}^n$ which vanish on V; and conversely to any ideal $\mathfrak{A} \subseteq {}_n\mathcal{O}_a$ there is associated a germ $\text{loc}\, \mathfrak{A}$ of a complex analytic subvariety at the point $a \in \mathbb{C}^n$, called the locus of the ideal $\mathfrak{A}$, on which all the functions in the ideal $\mathfrak{A}$ vanish. The detailed definitions and a further discussion of the properties of these operations can be found in CAV I; it suffices here merely to recall that $\text{loc id}\, V = V$ for any germ V of complex analytic subvariety and that $\text{id loc}\, \mathfrak{A} = \sqrt{\mathfrak{A}}$ for any ideal $\mathfrak{A} \subseteq {}_n\mathcal{O}_a$, where $\sqrt{\mathfrak{A}}$ denotes the radical of the ideal $\mathfrak{A}$. These operations consequently establish a one-to-one correspondence between germs of complex analytic subvarieties at a point $a \in \mathbb{C}^n$ and radical ideals in the local ring ${}_n\mathcal{O}_a$, where an ideal $\mathfrak{A} \subseteq {}_n\mathcal{O}_a$ is a radical ideal if $\mathfrak{A} = \sqrt{\mathfrak{A}}$; and thus the study of germs of complex analytic subvarieties at a point of $\mathbb{C}^n$ can be approached in a purely algebraic manner. A complex analytic homeomorphism φ from an open neighborhood of a point $a_1 \in \mathbb{C}^n$ to an open neighborhood of a point $a_2 \in \mathbb{C}^n$ induces in a familiar manner a ring isomorphism $\varphi^*: {}_n\mathcal{O}_{a_2} \longrightarrow {}_n\mathcal{O}_{a_1}$; and $\varphi^*(\text{id}\, \varphi(V)) = \text{id}\, V$ for any germ V of a complex analytic subvariety at a_1 and $\varphi(\text{loc}\, \varphi^*(\mathfrak{A})) = \text{loc}\, \mathfrak{A}$ for any ideal $\mathfrak{A} \subseteq {}_n\mathcal{O}_{a_2}$. Consequently there is a one-to-one correspondence between equivalence classes of germs of complex analytic subvarieties of $\mathbb{C}^n$ and equivalence classes of radical ideals in the local ring ${}_n\mathcal{O}_0$, where two ideals $\mathfrak{A}$, $\mathfrak{B}$ in ${}_n\mathcal{O}_0$ are equivalent if there is a complex analytic homeomorphism φ from an open neighborhood of the origin in $\mathbb{C}^n$ to another open neighborhood of the origin such that $\varphi(0) = 0$ and $\varphi^*(\mathfrak{A}) = \mathfrak{B}$; the problem of finding a purely algebraic description of these equivalence classes will be taken up in the next section.

To any germ V of a complex analytic subvariety at a point $a \in \mathbb{C}^n$ there is also associated the residue class ring ${}_V\mathcal{O}_a = {}_n\mathcal{O}_a/\mathrm{id}\, V$, the ring of germs of holomorphic functions on the germ V on the local ring of the germ V. The elements of ${}_V\mathcal{O}_a$ can be identified with the restrictions to V of germs of holomorphic functions at the point a in $\mathbb{C}^n$, and hence can be viewed as germs of continuous complex-valued functions at the point a on V. Any continuous mapping $\varphi: V_1 \to V_2$ from a germ V_1 of a complex analytic subvariety at a point $a_1 \in \mathbb{C}^{n_1}$ to a germ V_2 of a complex analytic subvariety at a point $a_2 \in \mathbb{C}^{n_2}$ induces in a familiar manner a homomorphism φ^* from the ring of germs of continuous complex-valued functions at the point a_2 on V_2 to the ring of germs of continuous complex-valued functions at the point a_1 on V_1; and the mapping φ is complex analytic precisely when $\varphi^*({}_{V_2}\mathcal{O}_{a_2}) \subseteq {}_{V_1}\mathcal{O}_{a_1}$, as demonstrated in Theorem 10 of CAV I. Thus the two germs V_1, V_2 are equivalent germs of complex analytic varieties precisely when there is a topological equivalence $\varphi: V_1 \to V_2$ which induces a ring isomorphism $\varphi^*: {}_{V_2}\mathcal{O}_{a_2} \to {}_{V_1}\mathcal{O}_{a_1}$, and a germ of a complex analytic variety can consequently be described as a germ of a topological space V together with a distinguished subring ${}_V\mathcal{O}$ of the ring of germs of continuous complex-valued functions on V; once again this criterion is rather a mixture of algebraic and topological properties, although both natural and useful, and the problem of finding a purely algebraic description of these equivalence classes as well will also be taken up in the next section. First though the global form for a germ of complex analytic variety should be introduced. A <u>complex analytic variety</u> is a Hausdorff topological space V endowed with a distinguished subsheaf ${}_V\mathcal{O}$ of the sheaf of germs of continuous complex-valued functions on V such that each point $a \in V$ the germ of the space V at a together with the stalk ${}_V\mathcal{O}_a$ is the germ of a complex analytic variety.

(b) The purely algebraic description of equivalence classes of germs of complex analytic subvarieties and of germs of complex analytic varieties requires slightly more than just the ring structure which has thus far primarily been considered. The ring ${}_V\mathcal{O}$ of germs of holomorphic functions on a germ V of a complex analytic variety V contains the subfield $\mathbb{C}$ of constant complex-valued functions as a canonical subring; thus ${}_V\mathcal{O}$ can be viewed as a ring and as a module over the subring $C \subseteq {}_V\mathcal{O}$, hence as an <u>algebra over the complex numbers with an identity element</u>. A complex analytic mapping $\varphi\colon V_1 \to V_2$ between two germs of complex analytic varieties induces a ring homomorphism $\varphi^*\colon {}_{V_2}\mathcal{O} \to {}_{V_1}\mathcal{O}$

which is the identity mapping between the canonical subrings of constant complex-valued functions; hence φ^* is actually an algebra homomorphism preserving the identities, and the converse assertion is also true as follows.

Theorem 1. If V_1, V_2 are germs of complex analytic subvarieties at respective points $a_1 \in \mathbb{C}^{n_1}$, $a_2 \in \mathbb{C}^{n_2}$ and if $\varphi^*: {}_{V_2}\mathcal{O}_{a_2} \to {}_{V_1}\mathcal{O}_{a_1}$ is a homomorphism of algebras over the complex numbers preserving the identities, then there is a unique complex analytic mapping $\varphi: V_1 \to V_2$ which induces the homomorphism φ^*.

Proof. Any ring homomorphism preserving the identities obviously takes units into units; and a $\mathbb{C}$-algebra homomorphism φ^* preserving the identities also takes nonunits into nonunits, that is, $\varphi^*({}_{V_2}\mathfrak{m}_{a_2}) \subseteq {}_{V_1}\mathfrak{m}_{a_1}$. To see this suppose that $f \in {}_{V_2}\mathfrak{m}_{a_2}$ but that $\varphi^*(f) \notin {}_{V_1}\mathfrak{m}_{a_1}$, hence that f is a germ of a holomorphic function vanishing at a_2 but $\varphi^*(f)$ is a germ of a holomorphic function having a nonzero complex value c at a_1; thus $f - c$ is a unit in ${}_{V_2}\mathcal{O}_{a_2}$ but $\varphi^*(f - c) = \varphi^*(f) - c$ is a nonunit in ${}_{V_1}\mathcal{O}_{a_1}$, which is impossible. Note further that actually $\varphi^*({}_{V_2}\mathfrak{m}^{\nu}_{a_2}) \subseteq {}_{V_1}\mathfrak{m}^{\nu}_{a_1}$ for any positive integer ν.

Now let w_i be the coordinate functions in $\mathbb{C}^{n_2}$ for $i = 1,\ldots,n_2$, and let $f_i = \varphi^*(w_i | V_2) \in {}_{V_1}\mathcal{O}_{a_1}$; and select any germs $F_i \in {}_{n_1}\mathcal{O}_{a_1}$ such that $F_i | V_1 = f_i$. Note that

$w_i|V_2 - w_i(a_2) \in {}_{V_2}\mathcal{W}_{a_2}$ and hence that $f_i - w_i(a_2) = \varphi^*(w_i|V_2 - w_i(a_2)) \in {}_{V_1}\mathcal{W}_{a_1}$; thus $F_i(a_1) = f_i(a_1) = w_i(a_2)$. The functions F_i can be taken as the coordinate functions of a complex analytic mapping Φ from an open neighborhood of a_1 in $\mathbb{C}^{n_1}$ into $\mathbb{C}^{n_2}$ such that $\Phi(a_1) = a_2$; and the proof will be concluded by showing that $\Phi(V_1) \subseteq V_2$ and that $\Phi|V_1 = \varphi$ induces the homomorphism φ^*. For any germ $f \in {}_{n_2}\mathcal{O}_{a_2}$ set $\tilde{\Phi}^*(f) = \Phi^*(f)|V_1 = (f \circ \Phi)|V_1$ and $\tilde{\varphi}^*(f) = \varphi^*(f|V_2)$; this defines two homomorphisms of $\mathbb{C}$-algebras $\tilde{\Phi}^*: {}_{n_2}\mathcal{O}_{a_2} \longrightarrow {}_{V_1}\mathcal{O}_{a_1}$ and $\tilde{\varphi}^*: {}_{n_2}\mathcal{O}_{a_2} \longrightarrow {}_{V_1}\mathcal{O}_{a_1}$. Note that $\tilde{\varphi}^*(w_i) = \varphi^*(w_i|V_2) = f_i = F_i|V_1 = \tilde{\Phi}^*(w_i)$, and consequently that the homomorphisms $\tilde{\Phi}^*$ and $\tilde{\varphi}^*$ agree on any polynomial in the coordinate functions w_i since both are homomorphisms of complex algebras. Then since any germ $f \in {}_{n_2}\mathcal{O}_{a_2}$ can be written in the form $f = f' + f''$, where f' is a polynomial in the coordinate functions w_i and $f'' \in {}_{n_2}\mathcal{W}^{\nu}_{a_2}$ for any given positive integer ν, it follows that

$$\tilde{\Phi}^*(f) - \tilde{\varphi}^*(f) = \tilde{\Phi}^*(f'') - \tilde{\varphi}^*(f'') \in {}_{V_1}\mathcal{W}^{\nu}_{a_1}$$

for any given positive integer ν, and hence that

$$\tilde{\Phi}^*(f) - \tilde{\varphi}^*(f) \in \bigcap_{\nu=1}^{\infty} {}_{V_1}\mathcal{W}^{\nu}_{a_1} ;$$

but since ${}_{V_1}\mathcal{O}_{a_1}$ is a noetherian local ring it follows from

Nakayama's lemma that $\bigcap_{\nu=1}^{\infty} {}_{V_1}\mathcal{M}_{a_1}^{\nu} = 0$, and therefore that $\tilde{\Phi}^* = \tilde{\varphi}^*$. By construction $\text{id}\, V_2 \subseteq \ker \tilde{\varphi}^*$, and hence $\text{id}\, V_2 \subseteq \ker \tilde{\Phi}^*$ as well. On the one hand then $0 = \tilde{\Phi}^*(f) = (f \circ \Phi)|V_1$ whenever $f \in \text{id}\, V_2 \subseteq {}_{n_2}\mathcal{O}_{a_2}$, or equivalently $f|\Phi(V_1) = 0$ whenever $f \in \text{id}\, V_2 \subseteq {}_{n_2}\mathcal{O}_{a_2}$, so that $\Phi(V_1) \subseteq V_2$; the restriction $\Phi|V_1 = \varphi$ is therefore a complex analytic mapping $\varphi\colon V_1 \to V_2$. On the other hand the homomorphisms $\tilde{\Phi}^* = \tilde{\varphi}^*$ can be viewed as determining homomorphisms from ${}_{V_2}\mathcal{O}_{a_2}$ into ${}_{V_1}\mathcal{O}_{a_1}$, since both vanish on $\text{id}\, V_2 \subseteq {}_{n_2}\mathcal{O}_{a_2}$; but the homomorphism determined by $\tilde{\Phi}^*$ is precisely that induced by φ while the homomorphism determined by $\tilde{\varphi}^*$ is just φ^*, hence φ^* is induced by φ. Since uniqueness is obvious, the proof is thereby concluded.

Two immediate consequences of this theorem merit stating explicitly, to complement the discussion in the preceding section.

Corollary 1 to Theorem 1. Equivalence classes of germs of complex analytic subvarieties of $\mathbb{C}^n$ are in one-to-one correspondence with equivalence of radical ideals in ${}_n\mathcal{O}$, where two ideals $\mathfrak{A}$, $\mathfrak{B}$ in ${}_n\mathcal{O}$ are equivalent if $\mathfrak{B} = \varphi^*(\mathfrak{A})$ for some automorphism $\varphi^*\colon {}_n\mathcal{O} \to {}_n\mathcal{O}$ of $\mathbb{C}$-algebras with identities.

Corollary 2 to Theorem 1. Two germs V_1, V_2 of complex analytic subvarieties of $\mathbb{C}^{n_1}$, $\mathbb{C}^{n_2}$ respectively are equivalent germs of complex analytic varieties if their local rings ${}_{V_1}\mathcal{O}$, ${}_{V_2}\mathcal{O}$ are isomorphic as $\mathbb{C}$-algebras with identities. Consequently germs of complex analytic varieties are in one-to-one correspondence with isomorphism classes of $\mathbb{C}$-algebras with identities of the form ${}_n\mathcal{O}/\mathfrak{A}$ where $\mathfrak{A}$ is a radical ideal in ${}_n\mathcal{O}$.

In view of these observations the study of germs of complex analytic subvarieties and varieties can be reduced to the purely algebraic study of the local algebras ${}_n\mathcal{O}$; this approach will not be pursued fully here, since the main interest in these lectures lies in the interrelations between algebraic, geometric, and analytic properties, but it is nonetheless a very useful tool to have at one's disposal. The algebraic approach also suggests considering from the beginning residue class algebras ${}_n\mathcal{O}/\mathfrak{A}$ for arbitrary ideals $\mathfrak{A} \subseteq {}_n\mathcal{O}$ and not just for radical ideals, which amounts to studying what are called generalized or nonreduced complex analytic varieties; again though this approach will not be followed here, since from some points of view it seems natural to view such residue class algebras as auxiliary structures on ordinary complex analytic varieties.

It should be noted before passing on to other topics that for Theorem 1 to hold it really is necessary to consider the local rings ${}_V\mathcal{O}$ as $\mathbb{C}$-algebras and not just as rings. For example the mapping $\varphi^*: {}_1\mathcal{O} \to {}_1\mathcal{O}$ which associates to any power series $f = \sum_{n=0}^{\infty} a_n z^n \in {}_1\mathcal{O}$ the power series $\varphi^*(f) = \sum_{n=0}^{\infty} \bar{a}_n z^n \in {}_1\mathcal{O}$, where $\bar{a}_n$ is the complex conjugate of a_n, is a well defined ring homomorphism but is not a homomorphism of $\mathbb{C}$-algebras and hence cannot be induced by a complex analytic mapping.

(c) A complex analytic mapping $\varphi\colon V_1 \to V_2$ between two germs of complex analytic varieties is a <u>finite analytic mapping</u> if $\varphi^{-1}(0) = 0$, where 0 as usual denotes the base point of a germ of complex analytic variety. Most of the mappings which arose in the discussion of the local parametrization theorem in CAV I, including the branched analytic coverings and the simple analytic mappings between irreducible germs, were finite analytic mappings; and the present discussion can be viewed as extending and completing that in the last two chapters of CAV I.

Actually the study of finite analytic mappings in general can be reduced to the study of the special finite analytic mappings which appeared in the discussion of the local parametrization theorem. Note first of all that for any complex analytic mapping $\varphi\colon V_1 \to V_2$ the germs V_1, V_2 of complex analytic varieties can be represented by germs V_1, V_2 of complex analytic subvarieties at the origin in $\mathbb{C}^{n+m} = \mathbb{C}^n \times \mathbb{C}^m$ and $\mathbb{C}^n$, respectively, in such a manner that φ is induced by the natural projection mapping $\mathbb{C}^n \times \mathbb{C}^m \to \mathbb{C}^n$. To see this, select any germs V_1, V_2 of complex analytic subvarieties at the origin in $\mathbb{C}^m, \mathbb{C}^n$ representing the given germs V_1, V_2, and any complex analytic mapping Φ from an open neighborhood of the origin in $\mathbb{C}^m$ to an open neighborhood of the origin in $\mathbb{C}^n$ such that $\Phi | V_1 = \varphi$. The complex analytic mapping taking a point z near the origin in $\mathbb{C}^m$ to the point $(\Phi(z), z)$ in $\mathbb{C}^n \times \mathbb{C}^m$ has a nonsingular Jacobian, hence imbeds an open neighborhood of the origin in $\mathbb{C}^m$ as a complex analytic submanifold of an open neighborhood of the origin in $\mathbb{C}^n \times \mathbb{C}^m$; and the image of the subvariety V_1 under this mapping is therefore a complex analytic subvariety of an open neighborhood of the origin in $\mathbb{C}^n \times \mathbb{C}^m$ which also represents the germ V_1, and in terms of this representation the mapping φ is induced by the desired projection. Now if φ is a finite analytic mapping and is so represented, then

$$V_1 \cap \{z \in \mathbb{C}^{n+m} | \ z_1 = \dots = z_n = 0\} = 0 ;$$

and it follows from Theorem 8 (b) of CAV I that, after possibly a change of coordinates in $\mathbb{C}^n$, the coordinates in $\mathbb{C}^{n+m} = \mathbb{C}^n \times \mathbb{C}^m$ form a regular system of coordinates for the ideal of each irreducible component of the germ V_1 of a complex analytic subvariety. The restriction of the complex analytic mapping φ to any irreducible component of the germ V_1 is then a partial projection in the representation of that component described by the local parametrization theorem. Thus by Corollary 6 to Theorem 5 of CAV I the image of a k-dimensional irreducible component V_1' of the germ V_1 is the germ $\varphi(V_1')$ of a k-dimensional irreducible complex analytic subvariety at the origin in $\mathbb{C}^n$; indeed for suitable representative subvarieties in some open neighborhoods of the origins the natural projection mappings from $\mathbb{C}^{n+m}$ and $\mathbb{C}^n$ to $\mathbb{C}^k$ induce branched analytic coverings $\pi: V_1' \to \mathbb{C}^k$ and $\pi': \varphi(V_1') \longrightarrow \mathbb{C}^k$ such that $\pi = \pi' \circ \varphi$.

To describe this more conveniently, define a _generalized branched analytic covering_ $\varphi: V_1 \to V_2$ to be a proper, light, complex analytic mapping from a complex analytic variety V_1 to a complex analytic variety V_2, such that there exist complex analytic subvarieties $D_1 \subset V_1$, $D_2 \subset V_2$ for which $D_1 = \varphi^{-1}(D_2)$, $V_1 - D_1$ and $V_2 - D_2$ are dense open subsets of V_1 and V_2 respectively, and the restriction $\varphi\colon V_1 - D_1 \longrightarrow V_2 - D_2$ is a complex analytic covering projection. A branched analytic covering as considered in CAV I is really just the special case of a generalized branched analytic covering in which V_2 is a regular analytic variety; generalized branched analytic coverings are very much like ordinary branched analytic coverings, particularly when $V_2 - D_2$ is connected, but there are rather obvious differences when $V_2 - D_2$ is not connected. In these terms the discussion in the preceding paragraph can be summarized as follows.

<u>Theorem 2</u>. If $\varphi\colon V_1 \to V_2$ is a finite analytic mapping between two germs of complex analytic varieties, then for any irreducible component V_1' of the germ V_1 the image $\varphi(V_1') \subseteq V_2$ is an irreducible germ of a complex analytic subvariety of V_2 and the restriction $\varphi | V_1'\colon V_1' \to \varphi(V_1')$ is a generalized branched analytic covering.

Any analytic mapping $\varphi\colon V_1 \to V_2$ between two germs of complex analytic varieties induces a homomorphism $\varphi^*\colon {}_{V_2}\mathcal{O} \to {}_{V_1}\mathcal{O}$ of $\mathbb{C}$-algebras with identities, and conversely as a consequence of Theorem 1 any homomorphism $\varphi^*\colon {}_{V_2}\mathcal{O} \to {}_{V_1}\mathcal{O}$ of $\mathbb{C}$-algebras with identities is induced by a unique complex analytic mapping $\varphi\colon V_1 \to V_2$; there then naturally arises the problem of characterizing those homomorphisms which correspond to finite analytic mappings. Before turning to this problem, though, a simple algebraic consequence of Theorem 2 should be mentioned.

<u>Corollary 1 to Theorem 2</u>. *If $\varphi\colon V_1 \to V_2$ is a finite analytic mapping between two germs of complex analytic varieties, then $\varphi(V_1) = V_2$ if and only if the induced homomorphism $\varphi^*\colon {}_{V_2}\mathcal{O} \to {}_{V_1}\mathcal{O}$ is injective.*

Proof. If $\varphi(V_1) \subset V_2$, then by Theorem 2 the image is actually a proper analytic subvariety of V_2; there is thus a nonzero element $f \in {}_{V_2}\mathcal{O}$ such that $f|\varphi(V_1) = 0$, hence such that $\varphi^*(f) = 0$, so that φ^* is not injective. Conversely if φ^* is not injective, there is a nonzero element $f \in {}_{V_2}\mathcal{O}$ such that $\varphi^*(f) = 0$, hence such that $f|\varphi(V_1) = 0$; thus $\varphi(V_1)$ is contained in the subvariety of V_2 defined by the vanishing of the function f, so that $\varphi(V_1) \subset V_2$.

Of course it is true for an arbitrary complex analytic mapping $\varphi\colon V_1 \to V_2$ that when $\varphi(V_1) = V_2$ then φ^* is injective, as is evident from the proof of the above corollary; but it is not true for an arbitrary complex analytic mapping $\varphi\colon V_1 \to V_2$ that when φ^* is injective then $\varphi(V_1) = V_2$, so the use of Theorem 2 in the proof of the above corollary is an essential one. For example, the germ at the origin of the complex analytic mapping $\varphi\colon \mathbb{C}^2 \to \mathbb{C}^2$ defined by $\varphi(z_1,z_2) = (z_1,z_1z_2)$ is not a surjective mapping, since points of the form $(0,z_2)$ cannot be contained in the image of φ if $z_2 \neq 0$; but the image of any open neighborhood of the origin does contain an open subset of $\mathbb{C}^2$, hence the induced homomorphism φ^* is necessarily injective.

Theorem 3(a). A complex analytic mapping $\varphi: V_1 \to V_2$ between two germs of complex analytic varieties is a finite analytic mapping if and only if every element of ${}_{V_1}\mathcal{O}$ is integral over the subring $\varphi^*({}_{V_2}\mathcal{O}) \subseteq {}_{V_1}\mathcal{O}$; indeed if φ is a finite analytic mapping then ${}_{V_1}\mathcal{O}$ is a finitely generated integral algebraic extension of the subring $\varphi^*({}_{V_2}\mathcal{O}) \subseteq {}_{V_1}\mathcal{O}$.

Proof. As noted above the given germs of complex analytic varieties can be represented by germs V_1, V_2 of complex analytic subvarieties at the origins in $\mathbb{C}^{n+m}$, $\mathbb{C}^n$, respectively, in such a manner that φ is induced by the natural projection mapping $\mathbb{C}^{n+m} \to \mathbb{C}^n$. If φ is a finite analytic mapping it can also be assumed, after possibly a change of coordinates in $\mathbb{C}^n$, that the coordinates in $\mathbb{C}^{n+m}$ form a regular system of coordinates for the ideal $\operatorname{id} V_1 \subseteq {}_{n+m}\mathcal{O}$; then as in the argument on pages 15-16 of CAV I the residue class ring ${}_{V_1}\mathcal{O} = {}_{n+m}\mathcal{O}/\operatorname{id} V_1$ is a finitely generated integral algebraic extension of the subring ${}_n\mathcal{O}/{}_n\mathcal{O} \cap \operatorname{id} V_1 = \varphi^*({}_n\mathcal{O}) = \varphi^*({}_{V_2}\mathcal{O})$. Conversely if every element of ${}_{V_1}\mathcal{O}$ is integral over the subring $\varphi^*({}_n\mathcal{O}) = \varphi^*({}_{V_2}\mathcal{O})$ then in particular the restrictions $z_j|V_1 \in {}_{V_1}\mathcal{O}$ of the coordinates in $\mathbb{C}^{n+m}$ are integral over $\varphi^*({}_n\mathcal{O})$ for $n+1 \leq j \leq n+m$; it then follows as usual that there are Weierstrass polynomials $p_j \in {}_n\mathcal{O}[z_j] \cap \operatorname{id} V_1$ for $n+1 \leq j \leq n+m$, hence that after possibly a change of coordinates in $\mathbb{C}^n$ the coordinates in $\mathbb{C}^{n+m}$ form a regular system of coordinates for the ideal $\operatorname{id} V_1 \subseteq {}_{n+m}\mathcal{O}$ and the mapping φ induced by the natural projection $\mathbb{C}^{n+m} \to \mathbb{C}^n$

is therefore necessarily a finite analytic mapping. That serves to conclude the proof of the theorem.

To rephrase this result rather more concisely note that any ring homomorphism $\varphi^*: {}_{V_2}\mathcal{O} \to {}_{V_1}\mathcal{O}$ can be viewed as exhibiting the ring ${}_{V_1}\mathcal{O}$ as a module over the ring ${}_{V_2}\mathcal{O}$. A ring homomormorphism $\varphi^*: {}_{V_2}\mathcal{O} \to {}_{V_1}\mathcal{O}$ is called a finite homomorphism if it exhibits ${}_{V_1}\mathcal{O}$ as a finitely generated module over the ring ${}_{V_2}\mathcal{O}$.

Theorem 3(b). A complex analytic mapping $\varphi: V_1 \to V_2$ between two germs of complex analytic varieties is a finite analytic mapping if and only if the induced ring homomorphism $\varphi^*: {}_{V_2}\mathcal{O} \to {}_{V_1}\mathcal{O}$ is a finite homomorphism. There is therefore a one-to-one correspondence between finite analytic mappings $\varphi: V_1 \to V_2$ and finite homomorphisms $\varphi^*: {}_{V_2}\mathcal{O} \to {}_{V_1}\mathcal{O}$ of $\mathbb{C}$-algebras with identities.

Proof. The first assertion is an immediate consequence of Theorem 3(a) and of the observation that a ring homomorphism $\varphi^*: {}_{V_2}\mathcal{O} \to {}_{V_1}\mathcal{O}$ is finite precisely when ${}_{V_1}\mathcal{O}$ is a finitely generated integral algebraic extension of the subring $\varphi^*({}_{V_2}\mathcal{O}) \subseteq {}_{V_1}\mathcal{O}$; and the second assertion then follows from an application of Theorem 1.

It is useful to observe that a somewhat more extensive form of finiteness also holds for finite analytic mappings. Recall that to any complex analytic mapping $\varphi: V_1 \to V_2$ between two complex analytic varieties and any analytic sheaf $\mathcal{S}$ over V_1 there is naturally associated an analytic sheaf $\varphi_*(\mathcal{S})$ over V_2, the direct image of the sheaf $\mathcal{S}$ under the mapping φ. For a branched analytic covering $\varphi: V_1 \to \mathbb{C}^k$ it was demonstrated in CAV I that the direct image sheaf $\varphi_*({}_{V_1}\mathcal{O})$ is actually a coherent analytic sheaf; and the same assertion holds for generalized branched analytic coverings as well. Coherence is really a local property, of course, so for the proof it suffices merely to consider a germ of a generalized branched analytic covering; and it is just as easy to prove slightly more at the same time.

<u>Theorem 4</u>. If $\varphi: V_1 \to V_2$ is a finite analytic mapping between two germs of complex analytic varieties then the direct image $\varphi_*(\mathcal{S})$ of any coherent analytic sheaf $\mathcal{S}$ over V_1 is a coherent analytic sheaf over V_2.

Proof. Again the given germs of complex analytic varieties can be represented by germs V_1, V_2 of complex analytic subvarieties at the origins in $\mathbb{C}^{m+n}$, $\mathbb{C}^n$, respectively, in such a manner that φ is induced by the natural projection mapping $\mathbb{C}^{n+m} \to \mathbb{C}^n$. Choose any germ W_1 of complex analytic subvariety at the origin in $\mathbb{C}^{n+m}$ such that $V_1 \subseteq W_1$ and that the natural projection mapping $\mathbb{C}^{n+m} \to \mathbb{C}^n$ also induces a branched analytic covering $\phi: W_1 \to \mathbb{C}^n$; for example, W_1 can be taken to be the germ of complex analytic subvariety defined by the subset $p_{n+1},\ldots,p_{n+m}$ of the first set of canonical equations for the ideal $\text{id}\, V_1 \subseteq {}_{n+m}\mathcal{O}$. If $\mathcal{S}$ is a coherent analytic sheaf over V_1 its trivial extension $\tilde{\mathcal{S}}$ to the variety W_1 is a coherent analytic sheaf over W_1, as noted on pages 78-80 of CAV I; and since evidently $\varphi_*(\mathcal{S}) = \Phi_*(\tilde{\mathcal{S}})|V_2$, then in order to prove the coherence of $\varphi_*(\mathcal{S})$ it suffices to prove the coherence of $\Phi_*(\tilde{\mathcal{S}})$, referring again to CAV I. Thus the proof of the theorem has been reduced to the proof of the assertion for the special case of a branched analytic covering $\Phi: W_1 \to \mathbb{C}^n$. If $\mathcal{J}$ is any coherent analytic sheaf over W_1 then in some open neighborhood of the origin in W_1 there is an exact sequence of analytic sheaves of the form

$$ {}_{W_1}\mathcal{O}^{r_1} \longrightarrow {}_{W_1}\mathcal{O}^{r} \longrightarrow \mathcal{J} \longrightarrow 0 . $$

Now the stalk at a point $p \in W_2$ of the direct image of any of these sheaves is just the direct sum of the stalks of that sheaf at the finitely many points $\Phi^{-1}(p) \subseteq W_1$; clearly then the direct images of these sheaves form an exact sequence of analytic sheaves

$$ \Phi_*({}_{W_1}\mathcal{O})^{r_1} \longrightarrow \Phi_*({}_{W_1}\mathcal{O})^{r} \longrightarrow \Phi_*(\mathcal{J}) \longrightarrow 0 . $$

Since the direct image sheaf $\Phi_*({}_{W_1}\mathcal{O})$ is a coherent analytic sheaf as a consequence of Theorem 19(b) of CAV I, it follows immediately that $\Phi_*(\mathcal{J})$ is also a coherent analytic sheaf, and that serves to conclude the proof of the theorem.

(d) A complex analytic mapping $\varphi: V_1 \to V_2$ between two germs of complex analytic varieties is completely characterized by the induced homomorphism $\varphi^*: {}_{V_2}\mathcal{O} \longrightarrow {}_{V_1}\mathcal{O}$ of $\mathbb{C}$-algebras with identities. The image of the maximal ideal ${}_{V_2}\mathfrak{m} \subset {}_{V_2}\mathcal{O}$ under this homomorphism is a subset $\varphi^*({}_{V_2}\mathfrak{m}) \subseteq {}_{V_1}\mathfrak{m}$ which generates an ideal in the ring ${}_{V_1}\mathcal{O}$ called the <u>characteristic</u> <u>ideal</u> of the mapping φ or of the homomorphism φ^*; this ideal will be denoted by ${}_{V_1}\mathcal{O} \cdot \varphi^*({}_{V_2}\mathfrak{m})$, where as customary the notation means the ideal consisting of all finite sums $\Sigma_i f_i\varphi^*(g_i)$ where $f_i \in {}_{V_1}\mathcal{O}$, $g_i \in {}_{V_2}\mathfrak{m}$. This ideal can also be viewed as the submodule of the ${}_{V_2}\mathcal{O}$-module ${}_{V_1}\mathcal{O}$ generated by the action of the maximal ideal ${}_{V_2}\mathfrak{m} \subset {}_{V_2}\mathcal{O}$ on the module ${}_{V_1}\mathcal{O}$, and when considered in this fashion as an ${}_{V_2}\mathcal{O}$-module will be called the <u>characteristic</u> <u>module</u> of the mapping φ or of the homomorphism φ^*. Which point of view to adopt depends on which of the algebras ${}_{V_1}\mathcal{O}$ or ${}_{V_2}\mathcal{O}$ is considered as primary; from either point of view the construction is a natural and useful one, particularly in that it furnishes a convenient coarser invariant of the analytic mapping than the full homomorphism φ^*. The present discussion will for the most part be limited to the characteristic ideal.

If the germ V_2 is represented by a germ of complex analytic subvariety at the origin in $\mathbb{C}^n$, then the mapping φ when viewed as a complex analytic mapping $\varphi: V_1 \to \mathbb{C}^n$ is given by n coordinate functions and the characteristic ideal of φ is evidently the ideal in ${}_{V_1}\mathcal{O}$ generated by these coordinate functions; conversely for any given proper ideal in ${}_{V_1}\mathcal{O}$ a set of n generators for that ideal can be viewed as the coordinate functions of a complex analytic mapping $\varphi: V_1 \to \mathbb{C}^n$ and the given ideal is the characteristic ideal of the mapping φ. Thus any proper ideal in ${}_{V_1}\mathcal{O}$ is the characteristic ideal of some complex analytic mapping from V_1 to another germ of complex analytic variety, but may very well be the characteristic ideal of a number of quite different mappings. The condition that a complex analytic mapping $\varphi: V_1 \to V_2$ be a finite analytic mapping can be expressed purely in terms of the characteristic ideal of that mapping.

Theorem 5. A complex analytic mapping $\varphi: V_1 \to V_2$ between two germs of complex analytic varieties is a finite analytic mapping if and only if its characteristic ideal $\mathfrak{A} = {}_{V_1}\mathcal{O} \cdot \varphi^*({}_{V_2}\mathfrak{m}') \subset {}_{V_1}\mathcal{O}$ satisfies any of the following equivalent conditions:

(a) $\operatorname{loc} \mathfrak{A} = 0$, the base point of V_1;

(b) $\sqrt{\mathfrak{A}} = {}_{V_1}\mathfrak{m}'$;

(c) ${}_{V_1}\mathfrak{m}'^{\,n} \subseteq \mathfrak{A} \subseteq {}_{V_1}\mathfrak{m}'$ for some positive integer n;

(d) ${}_{V_1}\mathcal{O}/\mathfrak{A}$ is a finite-dimensional complex vector space.

Proof. Since the complex analytic subvariety $\varphi^{-1}(0) \subseteq V_1$ is evidently the locus of the characteristic ideal $\mathfrak{A}$, it is an immediate consequence of the definition that φ is a finite analytic mapping precisely when $\operatorname{loc} \mathfrak{A} = 0$; thus to prove the theorem it suffices merely to prove the equivalence of the four listed conditions.

Firstly, that (a) and (b) are equivalent is an obvious consequence of the Hilbert zero theorem on the germ of complex analytic variety V_1. Secondly, if the ideal $\mathfrak{A} \subset {}_{V_1}\mathcal{O}$ satisfies (b) and f_i are finitely many generators of the maximal ideal ${}_{V_1}\mathfrak{m}'$

there are positive integers n_i such that $f_i^{n_i} \in \mathfrak{A}$; but any element $f \in {}_{V_1}\mathfrak{m}$ can be written in the form $f = \Sigma_i g_i f_i$ for some germs $g_i \in {}_{V_1}\mathcal{O}$, and if n is sufficiently large then each term in the multinomial expansion of the product of any n such expressions will involve a factor $f_i^{n_i}$ for some index i, so that ${}_{V_1}\mathfrak{m}^n \subseteq \mathfrak{A} \subseteq {}_{V_1}\mathfrak{m}$. Since clearly any ideal satisfying (c) also satisfies (b), it follows that (b) and (c) are equivalent. Finally note for any positive integer n that

$$ {}_{V_1}\mathcal{O} / {}_{V_1}\mathfrak{m}^n \cong {}_{V_1}\mathcal{O} / {}_{V_1}\mathfrak{m} \oplus {}_{V_1}\mathfrak{m} / {}_{V_1}\mathfrak{m}^2 \oplus \dots \oplus {}_{V_1}\mathfrak{m}^{n-1} / {}_{V_1}\mathfrak{m}^n $$

as ${}_{V_1}\mathcal{O}$ -modules; but each module ${}_{V_1}\mathcal{O} / {}_{V_1}\mathfrak{m}$ or ${}_{V_1}\mathfrak{m}^i / {}_{V_1}\mathfrak{m}^{i+1}$ is a finitely generated ${}_{V_1}\mathcal{O}$ -module on which the ideal ${}_{V_1}\mathfrak{m}$ acts trivially, hence is actually a finitely generated module over ${}_{V_1}\mathcal{O} / {}_{V_1}\mathfrak{m} \cong \mathbb{C}$, and therefore ${}_{V_1}\mathcal{O} / {}_{V_1}\mathfrak{m}^n$ is a finite-dimensional complex vector space. Then if the ideal $\mathfrak{A} \subset {}_{V_1}\mathcal{O}$ satisfies (c) it follows from this observation, in view of the natural injection ${}_{V_1}\mathcal{O} / \mathfrak{A} \longrightarrow {}_{V_1}\mathcal{O} / {}_{V_1}\mathfrak{m}^n$, that ${}_{V_1}\mathcal{O} / \mathfrak{A}$ is also a finite-dimensional complex vector space, hence that the ideal $\mathfrak{A}$ satisfies (d). Conversely if the ideal $\mathfrak{A} \subset {}_{V_1}\mathcal{O}$ satisfies (d) consider the descending chain of ${}_{V_1}\mathcal{O}$ -modules

$$ \mathcal{O}_{V_1} / \mathfrak{A} \supseteq ({}_{V_1}\mathfrak{m} + \mathfrak{A}) / \mathfrak{A} \supseteq ({}_{V_1}\mathfrak{m}^2 + \mathfrak{A}) / \mathfrak{A} \supseteq \dots ; $$

since these are finite-dimensional complex vector spaces the sequence is eventually stable, so that $({}_{V_1}\mathfrak{m}^n + \mathfrak{A})/\mathfrak{A} = ({}_{V_1}\mathfrak{m}^{n+1} + \mathfrak{A})/\mathfrak{A}$ for some positive integer n, and it then follows from Nakayama's lemma that ${}_{V_1}\mathfrak{m}^n \subseteq \mathfrak{A}$ and the ideal satisfies (c). Therefore (c) and (d) are equivalent, and the proof of the theorem is thereby concluded.

The dimension of the complex vector space ${}_{V_1}\mathcal{O}/\mathfrak{A}$ is an integer invariant associated to the characteristic ideal of a finite analytic mapping which has some further interesting properties.

<u>Theorem 6.</u> If $\varphi\colon V_1 \to V_2$ is a finite analytic mapping between two germs of complex analytic varieties with characteristic ideal $\mathfrak{A} \subset {}_{V_1}\mathcal{O}$, then the dimension of the complex vector space ${}_{V_1}\mathcal{O}/\mathfrak{A}$ is the minimal number of generators of ${}_{V_1}\mathcal{O}$ as an ${}_{V_2}\mathcal{O}$-module.

Proof. First let $f_1,\ldots,f_n$ be any elements of ${}_{V_1}\mathcal{O}$ which generate ${}_{V_1}\mathcal{O}$ as an ${}_{V_2}\mathcal{O}$-module, so that an arbitrary $f \in {}_{V_1}\mathcal{O}$ can be written in the form

$$f = \varphi^*(g_1)\cdot f_1 + \ldots + \varphi^*(g_n)\cdot f_n \tag{1}$$

for some germs $g_i \in {}_{V_2}\mathcal{O}$; then writing $g_i = c_i + g_i'$ where $c_i \in \mathbb{C}$ and $g_i' \in {}_{V_2}\mathfrak{m}$, it follows from (1) that

$$f - c_1f_1 - \ldots - c_nf_n = \varphi^*(g_1')\cdot f_1 + \ldots + \varphi^*(g_n')\cdot f_n \in \mathfrak{A} .$$

Thus the mapping which takes a vector $(c_1,\ldots,c_n) \in \mathbb{C}^n$ to the residue class in ${}_{V_1}\mathcal{O}/\mathfrak{m}$ of the element $c_1f_1 + \ldots + c_nf_n \in {}_{V_1}\mathcal{O}$ is a surjective linear mapping from $\mathbb{C}^n$ to ${}_{V_1}\mathcal{O}/\mathfrak{m}$, and consequently $\dim_{\mathbb{C}}({}_{V_1}\mathcal{O}/\mathfrak{m}) \leq n$; that is to say, $\dim_{\mathbb{C}}({}_{V_1}\mathcal{O}/\mathfrak{m})$ is less than or equal to the minimal number of generators of ${}_{V_1}\mathcal{O}$ as an ${}_{V_2}\mathcal{O}$-module. On the other hand let $d = \dim_{\mathbb{C}}({}_{V_1}\mathcal{O}/\mathfrak{m})$ and select any elements $f_1,\ldots,f_d$ of ${}_{V_1}\mathcal{O}$ which represent a basis for the complex vector space ${}_{V_1}\mathcal{O}/\mathfrak{m}$; thus an arbitrary $f \in {}_{V_1}\mathcal{O}$ can be written in the form

$$f = c_1f_1 + \ldots + c_df_d + g \tag{2}$$

where $c_i \in \mathbb{C}$ and $g \in \mathfrak{m}$. Now the elements $f_1,\ldots,f_d$ generate a submodule $\mathcal{A}$ of the ${}_{V_2}\mathcal{O}$-module ${}_{V_1}\mathcal{O}$, and it follows from (2) that

$${}_{V_1}\mathcal{O} = \mathcal{A} + \mathfrak{m} = \mathcal{A} + \varphi^*({}_{V_2}\mathfrak{m})\cdot{}_{V_1}\mathcal{O} ;$$

but then as a consequence of Nakayama's lemma ${}_{V_1}\mathcal{O} = \mathcal{A}$, so that ${}_{V_1}\mathcal{O}$ has d generators as an ${}_{V_2}\mathcal{O}$-module and therefore $\dim_{\mathbb{C}}({}_{V_1}\mathcal{O}/\mathfrak{m})$ is greater than or equal to the minimal number of generators of ${}_{V_1}\mathcal{O}$ as an ${}_{V_2}\mathcal{O}$-module. Combining these two parts, it follows that $\dim_{\mathbb{C}}({}_{V_1}\mathcal{O}/\mathfrak{m})$ is equal to the minimal number of generators of ${}_{V_1}\mathcal{O}$ as an ${}_{V_2}\mathcal{O}$-module, which was to be proved.

Corollary 1 to Theorem 6. A finite analytic mapping $\varphi\colon V_1 \to V_2$ between two germs of complex analytic varieties is an analytic equivalence between V_1 and its image $\varphi(V_1) \subseteq V_2$ if and only if the characteristic ideal of the mapping φ is equal to the maximal ideal ${}_{V_1}\mathfrak{m} \subset {}_{V_1}\mathcal{O}$.

Proof. If φ is an analytic equivalence between V_1 and $\varphi(V_1)$ then the induced homomorphism $\varphi^*: {}_{\varphi(V_1)}\mathcal{O} \longrightarrow {}_{V_1}\mathcal{O}$ is an isomorphism and it is quite obvious that the characteristic ideal of the mapping φ is the maximal ideal ${}_{V_1}\mathfrak{m} \subset {}_{V_1}\mathcal{O}$. On the other hand if the characteristic ideal of the mapping φ is the maximal ideal ${}_{V_1}\mathfrak{m} \subset {}_{V_1}\mathcal{O}$ then it follows from Theorem 6 that ${}_{V_1}\mathcal{O}$ has a single generator as an ${}_{V_2}\mathcal{O}$-module, hence that ${}_{V_1}\mathcal{O} \cong \varphi^*({}_{V_2}\mathcal{O})$; and recalling from Theorem 2 and its Corollary that $\varphi(V_1)$ is a germ of a complex analytic variety and that $\varphi^*({}_{V_2}\mathcal{O}) \cong {}_{\varphi(V_1)}\mathcal{O}$, it follows from Theorem 1 that φ is an analytic equivalence between V_1 and $\varphi(V_1)$, and the proof of the corollary is therewith concluded.

It is perhaps worth stating explicitly the following consequence of Theorem 5 and of Corollary 1 to Theorem 6, even though the proof is quite trivial.

<u>Corollary 2 to Theorem 6</u>. Any elements $f_1,\ldots,f_n$ in the maximal ideal ${}_V\mathfrak{m}$ of a germ V of a complex analytic variety which vanish simultaneously only at the base point of that germ are the coordinate functions of a finite analytic mapping $\varphi: V \to \mathbb{C}^n$; the image $\varphi(V)$ is the germ of a complex analytic subvariety at the origin in $\mathbb{C}^n$, and the germs V and $\varphi(V)$ are equivalent

germs of complex analytic varieties if and only if the functions $f_1,\ldots,f_n$ generate the entire maximal ideal ${}_V$.

Turning next to more geometrical properties, a finite analytic mapping $\varphi\colon V_1 \to V_2$ between two germs of complex analytic varieties is said to have <u>branching order</u> r if it can be represented by a generalized branched analytic covering $\varphi\colon V_1 \to V_2$ of r sheets. Note that this is not only just the condition that the finite analytic mapping can be represented by a generalized branched analytic covering, but moreover the requirement that the representative generalized branched analytic covering have the well defined number r of sheets; so if the associated unbranched covering does not lie over a connected space it must have the same number r of sheets over each connected component. If $\varphi\colon V_1 \to V_2$ is a surjective finite analytic mapping and V_1 is an irreducible germ then as a consequence of Theorem 2 the mapping φ necessarily has some branching order r; or if $\varphi\colon V_1 \to V_2$ is a finite analytic mapping for which V_1 is a pure dimensional germ, V_2 is an irreducible germ, and $\dim V_1 = \dim V_2$, then again the mapping φ has some branching order r. In general V_1 and V_2 need not be pure dimensional.

<u>Theorem 7</u>. If $\varphi\colon V_1 \to V_2$ is a finite analytic mapping of branching order r between two germs of complex analytic varieties and φ has characteristic ideal $\mathfrak{A} \subset {}_{V_1}\mathcal{O}$ then $\dim_{\mathbb{C}}({}_{V_1}\mathcal{O}/\mathfrak{A}) \geq r$, and $\dim_{\mathbb{C}}({}_{V_1}\mathcal{O}/\mathfrak{A}) = r$ if and only if ${}_{V_1}\mathcal{O}$ is a free ${}_{V_2}\mathcal{O}$-module.

Proof. Let $\varphi: V_1 \to V_2$ be a generalized branched analytic covering of r sheets representing the given germ of a complex analytic mapping. If $\dim_{\mathbb{C}}({}_{V_1}\mathcal{O}/\mathcal{M}) = d$ it follows from Theorem 6 that there are d germs $f_1,\ldots,f_d$ in ${}_{V_1}\mathcal{O}$ which generate ${}_{V_1}\mathcal{O}$ as an ${}_{V_2}\mathcal{O}$-module. Now the functions f_i can be viewed as sections of the direct image sheaf $\varphi_*({}_{V_1}\mathcal{O})$ in an open neighborhood of the base point $0 \in V_2$, and as such they generate an analytic subsheaf $\mathcal{S} \subseteq \varphi_*({}_{V_1}\mathcal{O})$ over that neighborhood; the stalks of these two sheaves coincide at the base point $0 \in V_2$, and since the direct image sheaf $\varphi_*({}_{V_1}\mathcal{O})$ is a coherent analytic sheaf as a consequence of Theorem 4, these two sheaves must then coincide in a full open neighborhood of the base point 0 in V_2. (To see this, merely observe that $\varphi_*({}_{V_1}\mathcal{O})$ is generated by a finite number of sections near 0, and that these sections lie in the subsheaf $\mathcal{S}$ at the point 0 and hence in a full open neighborhood of the point 0.) Thus the sections f_i furnish a surjective homomorphism of analytic sheaves ${}_{V_2}\mathcal{O}^d \to \varphi_*({}_{V_1}\mathcal{O})$; and letting $\mathcal{K} \subseteq {}_{V_2}\mathcal{O}^d$ be the kernel of this homomorphism there results the exact sequence of coherent analytic sheaves

$$0 \to \mathcal{K} \to {}_{V_2}\mathcal{O}^d \to \varphi_*({}_{V_1}\mathcal{O}) \to 0$$

over an open neighborhood of the base point 0 in V_2. At a point $p \in V_2$ over which the mapping φ is an unbranched analytic covering of r sheets it is evident that $\varphi_*({}_{V_1}\mathcal{O})_p \cong {}_{V_2}\mathcal{O}^r_p$; hence considering the exact sheaf sequence at that point it follows that

$d \geq r$. If $d = r$ it further follows from the exact sheaf sequence that $\mathcal{K}_p = 0$ at such a point p; and since $\mathcal{K}$ is a coherent analytic subsheaf of ${}_{V_2}\mathcal{O}^d$ and these points are dense in V_2 necessarily $\mathcal{K} = 0$. (Indeed $\mathcal{K}$ is generated by some sections of ${}_{V_2}\mathcal{O}^d$, so for each irreducible component of V_2 either $\mathcal{K}_p = 0$ for all points p belonging only to that component or $\mathcal{K}_p \neq 0$ for all points p belonging only to that component.) In particular, if $d = r$ then $\mathcal{K}_0 = 0$, and consequently ${}_{V_1}\mathcal{O}_0 = \varphi_*({}_{V_1}\mathcal{O})_0 \cong {}_{V_2}\mathcal{O}_0^r$; thus ${}_{V_1}\mathcal{O}$ is a free ${}_{V_2}\mathcal{O}$-module of rank r. On the other hand if ${}_{V_1}\mathcal{O}$ is a free ${}_{V_2}\mathcal{O}$-module it must be an ${}_{V_2}\mathcal{O}$-module of rank d, as a consequence of Theorem 6, and $\mathcal{K}_0 = 0$ in the exact sheaf sequence; thus $\mathcal{K} = 0$ and $\varphi_*({}_{V_1}\mathcal{O}) \cong {}_{V_2}\mathcal{O}^d$, so that again $d = r$. That suffices to complete the proof of the theorem.

One rather obvious special case of this theorem, which is nonetheless worth mentioning separately, is the following.

Corollary 1 to Theorem 7. If V is a germ of a complex analytic variety of pure dimension k and $f_1, \ldots, f_k$ are elements of the local ring ${}_V\mathcal{O}$ which generate an ideal $\mathfrak{A} \subset {}_V\mathcal{O}$ such that $\sqrt{\mathfrak{A}} = {}_V\mathfrak{m}$, then $f_1, \ldots, f_k$ are the coordinate functions of a branched analytic covering $\varphi: V \to \mathbb{C}^k$ of branching order r where $r \leq \dim_{\mathbb{C}}({}_V\mathcal{O}/\mathfrak{A}) < \infty$; and if $r = \dim_{\mathbb{C}}({}_V\mathcal{O}/\mathfrak{A})$ then the induced homomorphism $\varphi^*: {}_k\mathcal{O} \longrightarrow {}_V\mathcal{O}$ exhibits ${}_V\mathcal{O}$ as a free ${}_k\mathcal{O}$-module of rank r.

(e) The definitions of weakly holomorphic functions and of meromorphic functions on a complex analytic variety were given in CAV I, but the discussion of their properties was for the most part limited to the case of pure dimensional complex analytic varieties. The extension of that discussion to general complex analytic varieties is quite straightforward, but for completeness will be included here before turning to the consideration of the behavior of these classes of functions under finite analytic mappings.

The ring of germs of weakly holomorphic functions on a germ V of a complex analytic variety will be denoted by ${}_V\hat{\mathcal{O}}$, and the ring of germs of meromorphic functions on V will be denoted by ${}_V\mathcal{M}$, as before. Recall that a function $f \in {}_V\hat{\mathcal{O}}$ has a well defined value $f(0)$ at the base point $0 \in V$ if V is irreducible, although not in general (page 157 of CAV I); and that ${}_V\mathcal{M}$ is a field precisely when V is irreducible (page 136 of CAV I). An element $d \in {}_V\mathcal{O}$ is called a <u>universal denominator</u> for V if $d \cdot {}_V\hat{\mathcal{O}} \subseteq {}_V\mathcal{O}$; this is not the definition that was used in the pure-dimensional case in CAV I, but is evidently an equivalent definition in view of Corollary 1 to Theorem 24 and the discussion in §6(e) in CAV I.

<u>Theorem 8</u>. There exists a holomorphic function d in an open neighborhood of any point of a complex analytic variety such that d is a universal denominator but not a zero divisor at each point of that neighborhood.

Proof. Represent an open neighborhood of any point of the

given complex analytic variety by a complex analytic subvariety V of an open neighborhood U in $\mathbb{C}^n$, and write V as a union $V = \bigcup_i V_i$ of irreducible components. For each component V_i there exists after shrinking the neighborhood U if necessary a holomorphic function d_i which is a universal denominator for V_i but not a zero divisor on V_i at each point of V_i, as a consequence of Theorem 21 of CAV I; and $d_i = D_i|V_i$ for some function D_i holomorphic in U. There also exist after shrinking the neighborhood U if necessary holomorphic functions H_i in U such that $H_i|V_i \neq 0$ but $H_i|V_j = 0$ whenever $i \neq j$. The function $d = (\Sigma_i H_iD_i)|V$ is then holomorphic on V and is not a zero divisor at any point of V since it is nonzero on each irreducible component of V. If $f \in {}_V\hat{\mathcal{O}}_p$ at some point $p \in V$ then $f|V_i \in {}_{V_i}\hat{\mathcal{O}}_p$ whenever $p \in V_i$ so that $d_i\cdot(f|V_i) \in {}_{V_i}\mathcal{O}_p$; hence there is a germ $F_i \in {}_n\mathcal{O}_p$ such that $d_i\cdot(f|V_i) = F_i|V_i$. Note then that $F = \Sigma_i H_iF_i \in {}_n\mathcal{O}_p$ has the property that $F|V_i = (H_iF_i)|V_i = (H_id_if)|V_i = (H_iD_if)|V_i = (df)|V_i$, hence $df = F|V \in {}_V\mathcal{O}_p$; thus d is a universal denominator at any point $p \in V$, and the proof of the theorem is thereby concluded.

<u>Corollary 1 to Theorem 8</u>. On any complex analytic variety V the weakly holomorphic functions are precisely the locally bounded meromorphic functions; consequently at any point $p \in V$ the ring ${}_V\mathcal{M}_p$ of germs of meromorphic functions is also the total quotient ring of the ring ${}_V\hat{\mathcal{O}}_p$ of germs of weakly holomorphic functions.

Proof. It follows immediately from Theorem 8 that any weakly holomorphic function is meromorphic; and conversely any locally bounded meromorphic function is holomorphic at each regular point of V, as a consequence of the generalized Riemann removable singularities theorem, hence is actually weakly holomorphic. Since then ${}_V\mathcal{O}_p \subseteq {}_V\hat{\mathcal{O}}_p \subset {}_V\mathcal{M}_p$, and ${}_V\mathcal{M}_p$ is by definition the total quotient ring of ${}_V\mathcal{O}_p$, it follows that ${}_V\mathcal{M}_p$ is also the total quotient ring of ${}_V\hat{\mathcal{O}}_p$, and the proof of the corollary is thereby concluded.

<u>Corollary 2 to Theorem 8</u>. If $V = V_1 \cup \ldots \cup V_r$ is a germ of complex analytic variety with irreducible components V_i then

(a) ${}_V\hat{\mathcal{O}} \cong {}_{V_1}\hat{\mathcal{O}} \oplus \ldots \oplus {}_{V_r}\hat{\mathcal{O}}$,

(b) ${}_V\mathcal{M} \cong {}_{V_1}\mathcal{M} \oplus \ldots \oplus {}_{V_r}\mathcal{M}$.

Proof. If $f \in {}_V\hat{\mathcal{O}}$ then $f|V_i \in {}_{V_i}\hat{\mathcal{O}}$, and this yields an injective ring homomorphism ${}_V\hat{\mathcal{O}} \longrightarrow {}_{V_1}\hat{\mathcal{O}} \oplus \ldots \oplus {}_{V_r}\hat{\mathcal{O}}$; and since $V_i \cap V_j \subseteq \mathcal{S}(V)$ whenever $i \neq j$ it follows that for any elements $f_i \in {}_{V_i}\hat{\mathcal{O}}$ there is a well defined element $f \in {}_V\hat{\mathcal{O}}$ given by $f|V_i \cap \mathcal{R}(V) = f_i$, hence this homomorphism is actually an isomorphism and (a) thus holds rather trivially. Since ${}_V\mathcal{M}$ is the total quotient ring of ${}_V\hat{\mathcal{O}}$ as a consequence of Corollary 1 to Theorem 8 then (b) follows immediately from (a) and the proof of the corollary is thereby concluded.

<u>Corollary 3 to Theorem 8</u>. On any germ V of complex analytic variety the natural inclusion ${}_V\mathcal{O} \to {}_V\hat{\mathcal{O}}$ exhibits ${}_V\hat{\mathcal{O}}$ as a finitely generated ${}_V\mathcal{O}$-module; indeed ${}_V\hat{\mathcal{O}}$ is the integral closure of the ring ${}_V\mathcal{O}$ in its total quotient ring ${}_V\mathcal{M}_p$.

Proof. Since ${}_V\hat{\mathcal{O}} \cong {}_{V_1}\hat{\mathcal{O}} \oplus \ldots \oplus {}_{V_r}\hat{\mathcal{O}}$ where V has the decomposition into irreducible components $V = V_1 \cup \ldots \cup V_r$, as a consequence of Corollary 2 to Theorem 8, and ${}_{V_i}\hat{\mathcal{O}}$ is a finitely generated module over ${}_{V_i}\mathcal{O}$ hence also over ${}_V\mathcal{O}$, as a consequence of Corollary 2 to Theorem 24 of CAV I, it follows immediately that ${}_V\hat{\mathcal{O}}$ is a finitely generated ${}_V\mathcal{O}$-module. Any $f \in {}_V\hat{\mathcal{O}}$ does belong to ${}_V\mathcal{M}$, and since ${}_V\hat{\mathcal{O}}$ is a finite ${}_V\mathcal{O}$-module it follows as usual that f is integral over ${}_V\mathcal{O}$; conversely if $f \in {}_V\mathcal{M}$ is integral over ${}_V\mathcal{O}$ then its values whenever defined are the roots of a monic polynomial with holomorphic coefficients hence are locally bounded, so that $f \in {}_V\hat{\mathcal{O}}$ as a consequence of Corollary 1 to Theorem 8. Therefore ${}_V\hat{\mathcal{O}}$ is the integral closure of ${}_V\mathcal{O}$ in ${}_V\mathcal{M}$ and the proof of the corollary is thereby concluded.

Since the holomorphic functions on any open subset of a complex analytic variety V are a subring of the weakly holomorphic functions on that set, it is apparent that the sheaf of germs of weakly holomorphic functions has the natural structure of an analytic sheaf; this sheaf will also be denoted by ${}_V\hat{\mathcal{O}}$, since its stalk at any point $p \in V$ is the ${}_V\mathcal{O}_p$-module ${}_V\hat{\mathcal{O}}_p$.

Theorem 9. On any complex analytic variety V the sheaf ${}_V\hat{\mathcal{O}}$ of germs of weakly holomorphic functions is a coherent analytic sheaf.

Proof. Write the variety V as a union of pure-dimensional components $V = V_1 \cup \dots \cup V_r$ in an open neighborhood of any point. For each component V_i the sheaf ${}_{V_i}\hat{\mathcal{O}}$ of germs of weakly holomorphic functions is a coherent analytic sheaf over V_i, as discussed on page 159 of CAV I; and the trivial extension of this sheaf is then a coherent analytic sheaf over V in that neighborhood. Now the direct sum of these sheaves is also a coherent analytic sheaf over V, and since that direct sum coincides with the sheaf ${}_V\hat{\mathcal{O}}$ as a consequence of Corollary 2 to Theorem 8 it follows that ${}_V\hat{\mathcal{O}}$ is a coherent analytic sheaf over V, which was to be proved.

The set of all universal denominators at a point p of a complex analytic variety V clearly form a nontrivial ideal in the local ring ${}_V\mathcal{O}_p$; this ideal will be called the ideal of universal denominators for V at p or the conductor of V at p, and will be denoted by ${}_V\hat{\mathcal{J}}_p$. Note that ${}_V\hat{\mathcal{J}}_p \subseteq {}_V\mathcal{O}_p \subseteq {}_V\hat{\mathcal{O}}_p$, and that ${}_V\hat{\mathcal{J}}_p$ is also an ideal in ${}_V\hat{\mathcal{O}}_p$; indeed it is easy to see that ${}_V\hat{\mathcal{J}}_p$ can be characterized as the largest ideal in ${}_V\mathcal{O}_p$ which is also an ideal in ${}_V\hat{\mathcal{O}}_p$. (If $\mathfrak{A} \subseteq {}_V\mathcal{O}_p \subseteq {}_V\hat{\mathcal{O}}_p$ is an ideal in both rings then whenever $a \in \mathfrak{A}$ and $f \in {}_V\hat{\mathcal{O}}_p$ necessarily $af \in \mathfrak{A} \subseteq {}_V\mathcal{O}_p$, hence a is a universal denominator for V at p; thus $\mathfrak{A} \subseteq {}_V\hat{\mathcal{J}}_p$, and that demonstrates the assertion.) The set of all the conductors for V form a sheaf of ideals ${}_V\hat{\mathcal{J}} \subseteq {}_V\mathcal{O}$

over the complex analytic variety V.

<u>Corollary 1 to Theorem 9.</u> On any complex analytic variety V the sheaf of ideals ${}_V\mathcal{N}$ is a coherent analytic sheaf.

Proof. Since ${}_V\hat{\mathcal{O}}$ and ${}_V\mathcal{O}$ are coherent analytic sheaves by Theorem 9, and

$$ {}_V\mathcal{N}_p = \{f \in {}_V\mathcal{O}_p \mid f\cdot {}_V\hat{\mathcal{O}}_p \subseteq {}_V\mathcal{O}_p\} , $$

it follows immediately that ${}_V\mathcal{N}$ is a coherent analytic sheaf as well; the argument is quite standard, and can be found on page 142 of CAV I for example.

Since ${}_V\mathcal{N}$ is a coherent sheaf of ideals in ${}_V\mathcal{O}$ there is a complex analytic subvariety $\text{loc}\ {}_V\mathcal{N} \subset V$ such that the germ of that subvariety at any point $p \in V$ is the germ $\text{loc}\ {}_V\mathcal{N}_p$. Note that ${}_V\hat{\mathcal{O}}_p = {}_V\mathcal{O}_p$ at any regular point $p \in \mathcal{R}(V)$, so that ${}_V\mathcal{N}_p = {}_V\mathcal{O}_p$ whenever $p \in \mathcal{R}(V)$; this can be restated as follows.

<u>Corollary 2 to Theorem 9.</u> For any complex analytic variety $\text{loc}\ {}_V\mathcal{N} \subseteq \mathcal{S}(V)$.

Applying the Hilbert zero theorem it follows from Corollary 2 to Theorem 9 that at any point $p \in V$

$$ \sqrt{{}_V\mathcal{N}_p} = \text{id loc}\ {}_V\mathcal{N}_p \supseteq \text{id}\ \mathcal{S}(V)_p . $$

Consequently whenever $f \in {}_V\mathcal{O}_p$ vanishes on $\mathcal{S}(V)$ near the point p then some power of f belongs to the conductor ${}_V\mathcal{N}_p$;

this can be restated as follows.

<u>Corollary 3 to Theorem 9</u>. For any germ V of complex analytic variety and any function $f \in {}_V\mathcal{O}$ which vanishes on the singular locus $\mathcal{S}(V) \subset V$ there is a positive integer ν such that f^{ν} is a universal denominator for the germ V.

A germ of complex analytic variety V is said to be <u>normal</u> if ${}_V\hat{\mathcal{O}} = {}_V\mathcal{O}$, and correspondingly a complex analytic variety V is said to be normal at a point $p \in V$ if the germ of V at the point p is normal, hence if ${}_V\hat{\mathcal{O}}_p = {}_V\mathcal{O}_p$. The set of points at which a complex analytic variety is not normal is thus the complex analytic subvariety $\text{loc}\ {}_V\mathcal{J} \subset V$; and consequently if a variety is normal at a point p it is normal at all points of a full open neighborhood of the point p. Obviously a variety which is normal at a point p must be irreducible at the point p, indeed must actually be irreducible at all points of a full open neighborhood of the point p. For any irreducible germ V of complex analytic variety it was demonstrated in CAV I that there is a unique germ of complex analytic variety $\hat{V}$ such that ${}_{\hat{V}}\mathcal{O} \cong {}_V\hat{\mathcal{O}}$; indeed the isomorphism ${}_{\hat{V}}\mathcal{O} \cong {}_V\hat{\mathcal{O}}$ is induced by a simple analytic mapping $\rho: \hat{V} \to V$. The germ $\hat{V}$ is called the <u>normalization</u> of the germ V, and is itself a normal germ of complex analytic variety. For a reducible germ $V = V_1 \cup \dots \cup V_r$ of complex analytic variety with irreducible components V_i the normalization $\hat{V}$ is defined to be the disjoint union of the normalizations $\hat{V}_i$ of the components; again ${}_{\hat{V}}\mathcal{O} \cong {}_V\hat{\mathcal{O}}$, and the simple analytic mappings

$\rho_i: \hat{V}_i \to V_i$ can be viewed as forming a single simple analytic mapping $\rho: \hat{V} \to V$ inducing this isomorphism.

Theorem 10. If $\varphi: V_1 \to V_2$ is a finite analytic mapping of branching order r between two germs of complex analytic varieties then the homomorphisms induced by φ exhibit ${}_{V_1}\hat{\mathcal{O}}$ as a finitely generated integral algebraic extension of degree r of ${}_{V_2}\hat{\mathcal{O}}$ and ${}_{V_1}\mathcal{M}$ as an algebraic extension field of degree r over ${}_{V_2}\mathcal{M}$.

Proof. Since the given germ of a complex analytic mapping can be represented by a generalized branched analytic covering $\varphi: V_1 \to V_2$ it is evident that the induced homomorphisms $\varphi^*: {}_{V_2}\hat{\mathcal{O}} \to {}_{V_1}\hat{\mathcal{O}}$ and $\varphi^*: {}_{V_2}\mathcal{M} \to {}_{V_1}\mathcal{M}$ are well defined injective homomorphisms, hence can be viewed as exhibiting ${}_{V_2}\hat{\mathcal{O}}$ as a subring of ${}_{V_1}\hat{\mathcal{O}}$ and ${}_{V_2}\mathcal{M}$ as a subring of ${}_{V_1}\mathcal{M}$. (It should perhaps be noted for emphasis that if an analytic mapping $\varphi: V_1 \to V_2$ is not surjective then it does not necessarily induce well defined homomorphisms $\varphi^*: {}_{V_2}\hat{\mathcal{O}} \to {}_{V_1}\hat{\mathcal{O}}$ or $\varphi^*: {}_{V_2}\mathcal{M} \to {}_{V_1}\mathcal{M}$.) The composition of the normalization $\rho: \hat{V}_1 \to V_1$ and the mapping $\varphi: V_1 \to V_2$ is a finite analytic mapping from each irreducible component of $\hat{V}_1$ into V_2, so from Theorem 3(b) it follows readily that ${}_{V_1}\hat{\mathcal{O}} \cong {}_{\hat{V}_1}\mathcal{O}$ is a finite ${}_{V_2}\mathcal{O}$-module hence a finite ${}_{V_2}\hat{\mathcal{O}}$-module as well; thus ${}_{V_1}\hat{\mathcal{O}}$ is a finitely generated integral algebraic extension of the subring

${}_{V_2}\hat{\mathcal{O}}$. For the more precise result desired consider the associated unbranched analytic covering $\varphi: V_1 - D_1 \longrightarrow V_2 - D_2$, where D_i is an analytic subvariety of V_i and $V_i - D_i$ is a dense open subset of V_i for $i = 1,2$; for any point $z \in V_2 - D_2$ the inverse image $\varphi^{-1}(z)$ consists of r distinct points of $V_1 - D_1$ which in some order will be labeled $p_1(z),\ldots,p_r(z)$. If $f \in {}_{V_1}\hat{\mathcal{O}}$ then the polynomial $p_f(X) = \Pi_{i=1}^{r} (X - f(p_i(z)))$ is a monic polynomial of degree r in the variable X, and as in the proof of Theorem 18 in CAV I the coefficients are bounded holomorphic functions on $V_2 - D_2$ hence are elements of ${}_{V_2}\hat{\mathcal{O}}$; and since $p_f(f) = 0$ it follows that f is integral of degree r over ${}_{V_2}\hat{\mathcal{O}}$. Moreover if the values $f(p_i(z))$ are distinct for some point $z \in V_2 - D_2$ then f cannot be the root of any polynomial in ${}_{V_2}\hat{\mathcal{O}}[X]$ of degree strictly less that r. Thus ${}_{V_1}\hat{\mathcal{O}}$ is an integral algebraic extension of degree r of ${}_{V_2}\hat{\mathcal{O}}$. If $f \in {}_{V_1}\mathcal{M}$ the same argument shows that f is algebraic of degree at most r over ${}_{V_2}\mathcal{M}$, and that ${}_{V_1}\mathcal{M}$ is an algebraic extension of degree r of ${}_{V_2}\mathcal{M}$; and that suffices to conclude the proof of the theorem.

<u>Corollary 1 to Theorem 10</u>. If $\varphi: V_1 \to V_2$ is a finite analytic mapping of branching order r between two germs of complex analytic varieties with V_2 normal, and if $\mathfrak{A} \subset {}_{V_1}\mathcal{O}$ is the characteristic ideal of φ, then ${}_{V_1}\mathfrak{m}^r \subseteq \mathfrak{A} \subseteq {}_{V_1}\mathfrak{m}$.

Proof. As a consequence of Theorem 10 any element $f \in {}_{V_1}\mathcal{O}$ is the root of a monic polynomial $p_f(X) \in \varphi^*({}_{V_2}\hat{\mathcal{O}})[X]$, and the degree of $p_f(X)$ is at most r; if V_2 is normal then of course

${}_{V_2}\hat{\mathcal{O}} = {}_{V_2}\mathcal{O}$, and if $f \in {}_{V_1}\mathfrak{m} \subset {}_{V_1}\mathcal{O}$ then it is evident from the proof of Theorem 10 that all the coefficients of $p_f(X)$ except the leading coefficient actually belong to $\varphi^*({}_{V_2}\mathfrak{m})$. Thus

$$0 = p_f(f) = f^r + a_1 f^{r-1} + \dots + a_{r-1} f + a_r$$

where $a_i \in \varphi^*({}_{V_2}\mathfrak{m}) \subset {}_{V_1}\mathcal{O}$; but since $a_i f^{r-i} \in \varphi^*({}_{V_2}\mathfrak{m}) \cdot {}_{V_1}\mathcal{O} = \mathfrak{A}$ this shows that $f^r \in \mathfrak{A}$. Consequently $f^r \in \mathfrak{A}$ for every element $f \in {}_{V_1}\mathfrak{m}$. More generally for any elements $f_1, \dots, f_r$ of ${}_{V_1}\mathfrak{m}$ and any constants $c_1, \dots, c_r$ it follows that $(c_1 f_1 + \dots + c_r f_r)^r \in \mathfrak{A}$; and since this holds for arbitrary constants c_i, it follows clearly that each term in the multinomial expansion of this power must also belong to $\mathfrak{A}$, and consequently $f_1 \cdots f_r \in \mathfrak{A}$. That shows that ${}_{V_1}\mathfrak{m}^r \subseteq \mathfrak{A} \subseteq {}_{V_1}\mathfrak{m}$, and concludes the proof of the corollary.

Corollary 2 to Theorem 10. If V is a germ of a complex analytic variety of pure dimension k and $f_1, \dots, f_k$ are the coordinate functions of a branched analytic covering $\varphi: V \to \mathbb{C}^k$ of r sheets, then the germs $f_i \in {}_V\mathcal{O}$ generate an ideal $\mathfrak{A} \subset {}_V\mathcal{O}$ for which ${}_V\mathfrak{m}^r \subseteq \mathfrak{A} \subseteq {}_V\mathfrak{m}$.

Proof. This is of course just the special case of Corollary 1 to Theorem 10 in which $V_2 = \mathbb{C}^k$.

§2. Finite analytic mappings with given domain

(a) Consider the problem of describing all finite analytic mappings from a given germ V of a complex analytic variety into another germ of complex analytic variety. The image of any such mapping is itself a germ of a complex analytic variety as a consequence of Theorem 2, so the mapping can be viewed as the composition of a surjective finite analytic mapping and an inclusion mapping; and the present interest centers on describing only the first of these two factors. If $\varphi: V \to W$ is a surjective finite analytic mapping then the induced homomorphism $\varphi^*: {}_W\mathcal{O} \to {}_V\mathcal{O}$ is injective by Corollary 1 to Theorem 2, so the image $\varphi^*({}_W\mathcal{O}) = \mathcal{R}$ is a subalgebra of ${}_V\mathcal{O}$ isomorphic to ${}_W\mathcal{O}$. Conversely given a subalgebra $\mathcal{R} \subseteq {}_V\mathcal{O}$ and an isomorphism $\varphi^*: {}_W\mathcal{O} \to \mathcal{R}$ for some germ W of complex analytic variety it follows from Theorem 1 that there is a complex analytic mapping $\varphi: V \to W$ inducing the homomorphism φ^*; and φ is a finite analytic mapping precisely when its characteristic ideal, which can be described in terms of the subalgebra $\mathcal{R}$, satisfies one of the conditions in Theorem 5. This provides a purely algebraic approach to the problem of interest here, but is still rather unsatisfactory in that the description of the subalgebra $\mathcal{R} \subseteq {}_V\mathcal{O}$ requires the existence of an isomorphism $\varphi^*: {}_W\mathcal{O} \to \mathcal{R}$ for some germ W of complex analytic variety; however this objection is easily overcome as follows.

Theorem 11. For any germ V of complex analytic variety, a subalgebra with identity $\mathcal{R} \subseteq {}_V\mathcal{O}$ is the image of the homomorphism induced by a finite analytic mapping from V to another germ of complex analytic variety if and only if the subalgebra $\mathcal{R}$ satisfies both of the following conditions:

(a) the ideal $\mathfrak{A} = {}_V\mathcal{O}\cdot(\mathcal{R} \cap {}_V\mathfrak{m})$ in ${}_V\mathcal{O}$ generated by the ideal $\mathcal{R} \cap {}_V\mathfrak{m}$ in $\mathcal{R}$ has the property that $\sqrt{\mathfrak{A}} = {}_V\mathfrak{m}$; and

(b) $\mathcal{R} = \bigcap_{\nu=1}^{\infty} (\mathcal{R} + {}_V\mathfrak{m}^{\nu})$.

Proof. First suppose that $\varphi\colon V \to W$ is a finite analytic mapping, which can of course be assumed surjective, and that $\mathcal{R} = \varphi^*({}_W\mathcal{O}) \subseteq {}_V\mathcal{O}$. The characteristic ideal of the mapping φ is then the ideal ${}_V\mathcal{O}\cdot\varphi^*({}_W\mathfrak{m}) = {}_V\mathcal{O}\cdot({}_V\mathfrak{m} \cap \varphi^*({}_W\mathcal{O})) = \mathfrak{A}$, and it follows from Theorem 5 that $\sqrt{\mathfrak{A}} = {}_V\mathfrak{m}$ so that condition (a) is necessarily satisfied. It also follows from Theorem 5 that ${}_V\mathfrak{m}^n \subseteq \mathfrak{A} \subseteq {}_V\mathfrak{m}$ for some positive integer n, so that in order to prove that condition (b) is satisfied it suffices to show that $\mathcal{R} = \mathcal{S}$ where

$$\mathcal{S} = \bigcap_{\nu=1}^{\infty} (\mathcal{R} + \mathfrak{A}^{\nu}) = \bigcap_{\nu=1}^{\infty} (\mathcal{R} + \mathfrak{m}^{\nu}\cdot{}_V\mathcal{O})$$

and $\mathfrak{m} = \varphi^*({}_W\mathfrak{m})$; here $\mathcal{R}$ is isomorphic to ${}_W\mathcal{O}$ and $\mathfrak{m}$ corresponds to ${}_W\mathfrak{m}$ under this isomorphism. Note that $\mathcal{R} \subseteq \mathcal{S} \subseteq {}_V\mathcal{O}$, so that $\mathcal{S}$ and ${}_V\mathcal{O}$ can be viewed as $\mathcal{R}$-modules; and since ${}_V\mathcal{O}$ is a finite $\mathcal{R}$-module as a consequence of Theorem 3(b) and

$\mathcal{R} \cong {}_W\mathcal{O}$ is a Noetherian ring, it follows that $\mathcal{S}$ is a finite $\mathcal{R}$-module as well. Passing to the quotient modules $\tilde{\mathcal{S}} = \mathcal{S}/\mathcal{R}$ and ${}_V\tilde{\mathcal{O}} = {}_V\mathcal{O}/\mathcal{R}$, which can also be viewed as finite $\mathcal{R}$-modules, and observing that $\tilde{\mathcal{S}} = \bigcap_{\nu=1}^{\infty} \mathfrak{m}^{\nu}\cdot{}_V\tilde{\mathcal{O}}$, it also follows that $\tilde{\mathcal{S}} = \mathfrak{m}\cdot\tilde{\mathcal{S}}$; but then by Nakayama's lemma $\tilde{\mathcal{S}} = 0$, hence $\mathcal{S} = \mathcal{R}$ so condition (b) is also necessarily satisfied.

Conversely suppose that $\mathcal{R} \subseteq {}_V\mathcal{O}$ is a subalgebra with identity and that $\mathcal{R}$ satisfies conditions (a) and (b). The ideal $\mathcal{U} = {}_V\mathcal{O}\cdot(\mathcal{R} \cap {}_V\mathfrak{m}) \subseteq {}_V\mathcal{O}$ is generated by finitely many of its elements, say $f_1,\ldots,f_n$, and these can be chosen so that $f_i \in \mathcal{R} \cap {}_V\mathfrak{m}$. Since $\sqrt{\mathcal{U}} = {}_V\mathfrak{m}$ by condition (a), it follows from Theorem 5 that the functions f_i can be taken to be the coordinate functions of a finite analytic mapping $\psi\colon V \to \mathbb{C}^n$; and by Theorem 2 the image $\psi(V) = W_o$ is the germ of a complex analytic subvariety at the origin in $\mathbb{C}^n$. Note that any element $f \in {}_{W_o}\mathcal{O}$ can be written $f = f' + f''$ where f' is the restriction to W_o of a polynomial in the coordinate functions z_i in $\mathbb{C}^n$ and $f'' \in {}_{W_o}\mathfrak{m}^{\nu}$ for a given positive integer ν; and consequently $\psi^*(f) = \psi^*(f') + \psi^*(f'') \in \mathcal{R} + {}_V\mathfrak{m}^{\nu}$, since $\psi^*(z_i|W_o) = f_i \in \mathcal{R}$ and $\psi^*({}_{W_o}\mathfrak{m}) \subseteq {}_V\mathfrak{m}$. Thus $\psi^*({}_{W_o}\mathcal{O}) \subseteq \mathcal{R} + {}_V\mathfrak{m}^{\nu}$ for any positive integer ν, so it follows from condition (b) that $\psi^*({}_{W_o}\mathcal{O}) \subseteq \mathcal{R}$; and therefore $\mathcal{R}$ and ${}_V\mathcal{O}$ can be viewed as modules over the common subring $\psi^*({}_{W_o}\mathcal{O})$. Now ${}_V\mathcal{O}$ is a finite module over $\psi^*({}_{W_o}\mathcal{O})$ since ψ is a finite analytic mapping; and since $\psi^*({}_{W_o}\mathcal{O}) \cong {}_{W_o}\mathcal{O}$ is Noetherian then $\mathcal{R}$ is also a finite module over $\psi^*({}_{W_0}\mathcal{O})$. Choose elements $f_{n+1},\dots,f_{n+m}$ in $\mathcal{R} \cap {}_V\mathfrak{m}$ such that these elements together with the identity element $1 \in {}_V\mathcal{O}$ generate $\mathcal{R}$ as a module over $\psi^*({}_{W_0}\mathcal{O})$. All the elements $f_1,\dots,f_{n+m}$ together also generate the ideal $\mathfrak{A}$ in ${}_V\mathcal{O}$, so they too can be taken to be the coordinate functions of a finite analytic mapping $\varphi\colon V \to \mathbb{C}^{n+m}$; and just as before the image $\varphi(V) = W$ is the germ of a complex analytic subvariety at the origin in $\mathbb{C}^{n+m}$, and $\varphi^*({}_W\mathcal{O}) \subseteq \mathcal{R}$. However the mapping $\psi\colon V \to \mathbb{C}^n$ can be written as the composition of the mapping $\varphi\colon V \to \mathbb{C}^{n+m}$ and the natural projection mapping $\mathbb{C}^{n+m} \to \mathbb{C}^n$, so that $\psi^*({}_{W_0}\mathcal{O}) \subseteq \varphi^*({}_W\mathcal{O})$; and since any element $f \in \mathcal{R}$ can be written

$$f = g_0 \cdot 1 + g_1 \cdot f_{n+1} + \dots + g_m f_{n+m}$$

for some $g_i \in \psi^*({}_{W_0}\mathcal{O}) \subseteq \varphi^*({}_W\mathcal{O})$, and $f_{n+i} \in \varphi^*({}_W\mathcal{O})$ as well, it follows that $f \in \varphi^*({}_W\mathcal{O})$. Therefore $\mathcal{R} = \varphi^*({}_W\mathcal{O})$, and that suffices to conclude the proof of the theorem.

It should perhaps be noted that a completeness condition such as condition (b) is really necessary in the preceding theorem; for example the subalgebra $\mathbb{C}[z_1,\dots,z_n] \subset {}_n\mathcal{O}$ satisfies condition (a) but not condition (b), so does not correspond to a finite analytic mapping from the origin in $\mathbb{C}^n$ to any other germ of complex analytic variety. It should perhaps also be noted that in the last part of the proof of the preceding theorem it really is necessary to consider the subvariety W as well as the subvariety W_0, since it is not necessarily the case that $\psi^*({}_{W_0}\mathcal{O}) = \mathcal{R}$; that merel, reflects the fact that the characteristic ideal $\mathfrak{A}$ does not determine the subalgebra $\mathcal{R}$, as will be illustrated in the examples discussed in §3(b). It is quite easy though to determine whether a set of generators of the ideal $\mathfrak{A}$ are the coordinate functions of a finite analytic mapping $\phi: V \to W$ for which $\phi^*({}_W\mathcal{O}) = \mathcal{R}$. In this connection observe that whenever $\mathcal{R} \subseteq {}_V\mathcal{O}$ is a subalgebra with identity and $\mathcal{R}$ satisfies conditions (a) and (b), so that $\mathcal{R} = \phi^*({}_W\mathcal{O}) \cong {}_W\mathcal{O}$ for some surjective finite analytic mapping $\phi: V \to W$, then the maximal ideal of the local ring $\mathcal{R}$ is just the ideal ${}_{\mathcal{R}}\mathfrak{m} = \mathcal{R} \cap {}_V\mathfrak{m}$ and the characteristic ideal of the mapping ϕ is the ideal $\mathfrak{A} = {}_V\mathcal{O} \cdot {}_{\mathcal{R}}\mathfrak{m}$; the relation between the finite dimensional complex vector space ${}_{\mathcal{R}}\mathfrak{m} / {}_{\mathcal{R}}\mathfrak{m}^2$ and the imbedding dimension of the germ of complex analytic variety W with local ring ${}_W\mathcal{O} \cong \mathcal{R}$ was discussed in CAV I.

Corollary 1 to Theorem 11. Let V be a germ of complex analytic variety and $\mathcal{R} \subseteq {}_V\mathcal{O}$ be a subalgebra with identity such that $\mathcal{R}$ satisfies conditions (a) and (b) of Theorem 11. If $f_1, \ldots, f_r$ are any elements of ${}_{\mathcal{R}}\mathfrak{m} = \mathcal{R} \cap {}_V\mathfrak{m}$ which represent generators of the complex vector space ${}_{\mathcal{R}}\mathfrak{m} / {}_{\mathcal{R}}\mathfrak{m}^2$, then these functions are the coordinate functions of a finite analytic mapping $\phi: V \to \mathbb{C}^r$ with image $W = \phi(V) \subseteq \mathbb{C}^r$ and the induced homomorphism $\phi^*: {}_W\mathcal{O} \to {}_V\mathcal{O}$ is an isomorphism between ${}_W\mathcal{O}$ and $\mathcal{R}$.

Proof. It follows from Theorem 11 that there exists a surjective finite analytic mapping $\psi: V \to W_o$ such that the induced homomorphism $\psi^*: {}_{W_o}\mathcal{O} \to {}_V\mathcal{O}$ is an isomorphism between

${}_{W_0}\mathcal{O}$ and $\mathcal{R}$. The elements $f_i \in \mathcal{R} = \psi^*({}_{W_0}\mathcal{O})$ are therefore the images $f_i = \psi^*(g_i)$ of some elements $g_i \in {}_{W_0}\mathfrak{m}$, and the elements g_i in turn represent generators of the complex vector space ${}_{W_0}\mathfrak{m}/{}_{W_0}\mathfrak{m}^2$ and hence by Nakayama's lemma are also generators of the maximal ideal ${}_{W_0}\mathfrak{m} \subseteq {}_{W_0}\mathcal{O}$. It then follows from Corollary 2 to Theorem 6 that the functions g_i are the coordinate functions of a finite analytic mapping $\theta: W_0 \to \mathbb{C}^r$ which is an equivalence between the germs of complex analytic varieties W_0 and $W = \theta(W_0) \subseteq \mathbb{C}^r$. The functions $f_i = g_i \circ \psi$ are the coordinate functions of the finite analytic mapping $\phi = \theta \circ \psi$: $V \to W$, and $\phi^* = \psi^* \circ \theta^*$: ${}_W\mathcal{O} \longrightarrow {}_V\mathcal{O}$ is therefore an isomorphism between ${}_W\mathcal{O}$ and $\mathcal{R}$ as desired. That completes the proof of the corollary.

A special case of Theorem 11 that is of some interest is the characterization of simple analytic mappings. Recall that a simple analytic mapping between two germs of complex analytic varieties is a finite analytic mapping of branching order one; these mappings arise most naturally for pure-dimensional varieties.

Corollary 2 to Theorem 11. For a pure-dimensional germ V of complex analytic variety, a subalgebra with identity $\mathcal{R} \subseteq {}_V\mathcal{O}$ is the image of the homomorphism induced by a simple analytic mapping from V to another germ of complex analytic variety if and only if the subalgebra $\mathcal{R}$ satisfies conditions (a) and (b) of Theorem 11 and in addition

(c) $d \cdot {}_V\mathcal{O} \subseteq \mathcal{R}$ for some element $d \in {}_V\mathcal{O}$ which is not a zero divisor in ${}_V\mathcal{O}$.

Proof. A simple analytic mapping $\varphi: V \to W$ is a finite analytic mapping, hence the subalgebra $\mathcal{R} = \varphi^*({}_W\mathcal{O}) \subseteq {}_V\mathcal{O}$ must necessarily satisfy conditions (a) and (b) of Theorem 11. In addition it follows from Theorem 21 of CAV I that there is an element $g \in {}_W\mathcal{O}$ which is not a zero divisor in ${}_W\mathcal{O}$, and which is a relative denominator for the simple analytic mapping $\varphi: V \to W$, hence for which $\varphi^*(g)\cdot{}_V\mathcal{O} \subseteq \varphi^*({}_W\mathcal{O}) = \mathcal{R}$; but then $d = \varphi^*(g) \in {}_V\mathcal{O}$ clearly is not a zero divisor in ${}_V\mathcal{O}$ and $d\cdot{}_V\mathcal{O} \subseteq \mathcal{R}$, so that the subalgebra $\mathcal{R}$ necessarily satisfies condition (c) as well.

Conversely if $\mathcal{R} \subseteq {}_V\mathcal{O}$ is a subalgebra with identity and $\mathcal{R}$ satisfies conditions (a), (b), and (c), then it follows from Theorem 11 that there is a surjective finite analytic mapping $\varphi: V \to W$ such that $\mathcal{R} = \varphi^*({}_W\mathcal{O})$. The restriction of φ to each irreducible component of V can by Theorem 2 be represented by a generalized branched analytic covering, hence φ itself can be represented by a generalized branched analytic covering $\varphi: V \to W$; thus there are analytic subvarieties $D_1 \subset V$, $D_2 \subset W$ such that $V - D_1$ and $W - D_2$ are dense open subsets of V and W respectively and that the restriction $\varphi: V - D_1 \longrightarrow W - D_2$ is an unbranched analytic covering. If this is a covering of $r > 1$ sheets over some connected component of $W - D_2$ then there is an element $f \in {}_V\mathcal{O}$ which separates these r sheets, and since d cannot vanish identically on any connected component of $V - D_2$ it is obvious that $d\cdot f \notin \mathcal{R} = \varphi^*({}_W\mathcal{O})$, contradicting (c). Thus $\varphi: V - D_1 \longrightarrow W - D_2$ is a one-sheeted covering of each connected component, and φ is consequently a simple analytic mapping; that then suffices to conclude the proof.

The normalization of any irreducible germ of complex analytic variety gives a simple analytic mapping to which Corollary 2 can be applied to yield the following result.

Corollary 3 to Theorem 11. If V is a normal germ of complex analytic variety, then the set of germs of complex analytic varieties having normalization V is in one-to-one correspondence with the set of equivalence classes of subalgebras with identities $\mathcal{R} \subseteq {}_V\mathcal{O}$ satisfying conditions (a), (b), and (c) of Theorem 11 and its Corollary 2, where two subalgebras $\mathcal{R}_1, \mathcal{R}_2$ of ${}_V\mathcal{O}$ are equivalent if there is an algebra automorphism $\theta: {}_V\mathcal{O} \to {}_V\mathcal{O}$ such that $\theta(\mathcal{R}_1) = \mathcal{R}_2$.

Proof. It follows from Corollary 2 to Theorem 11 that the set of subalgebras $\mathcal{R} \subseteq {}_V\mathcal{O}$ satisfying conditions (a), (b), and (c) is in one-to-one correspondence with the set of simple analytic mappings from V to another germ of complex analytic variety, the correspondence being that which associates to a simple analytic mapping φ the subalgebra $\varphi^*({}_{\varphi(V)}\mathcal{O}) \subseteq {}_V\mathcal{O}$. Any simple analytic mapping $\varphi: V \to W$ exhibits the normal germ of complex analytic variety V as the normalization of the germ W, and of course any germ W having normalization V is the image of some simple analytic mapping $\varphi: V \to W$; and in addition to the homomorphism $\varphi^*: {}_W\mathcal{O} \to {}_V\mathcal{O}$ the mapping φ induces an isomorphism

$\hat{\varphi}: {}_W\hat{\mathcal{O}} \longrightarrow {}_V\hat{\mathcal{O}} = {}_V\mathcal{O}$ for which $\hat{\varphi}|_W\mathcal{O} = \varphi^*$. Now if $\varphi_1: V \rightarrow W_1$ and $\varphi_2: V \rightarrow W_2$ are simple analytic mappings for which the images W_1 and W_2 are equivalent germs of complex analytic varieties under an equivalence $\psi: W_1 \rightarrow W_2$, then ψ induces an algebra isomorphism $\hat{\psi}: {}_{W_2}\hat{\mathcal{O}} \longrightarrow {}_{W_1}\hat{\mathcal{O}}$; but then $\theta = \hat{\varphi}_1\hat{\psi}\hat{\varphi}_2^{-1}: {}_V\mathcal{O} \rightarrow {}_V\mathcal{O}$ is also an algebra isomorphism and $\theta(\varphi_2^*({}_{W_2}\mathcal{O})) = \hat{\varphi}_1\hat{\psi}({}_{W_2}\mathcal{O}) = \hat{\varphi}_1({}_{W_1}\mathcal{O}) = \varphi^*({}_{W_1}\mathcal{O})$. Conversely if $\varphi_1: V \rightarrow W_1$ and $\varphi_2: V \rightarrow W_2$ are simple analytic mappings for which there is an automorphism $\theta: {}_V\mathcal{O} \rightarrow {}_V\mathcal{O}$ such that $\theta(\varphi_2^*({}_{W_2}\mathcal{O})) = \varphi_1^*({}_{W_1}\mathcal{O})$, then the algebras ${}_{W_1}\mathcal{O}$ and ${}_{W_2}\mathcal{O}$ are isomorphic; hence by Corollary 2 to Theorem 1 the germs W_1 and W_2 are equivalent germs of complex analytic varieties. That suffices then to conclude the proof of the corollary.

There are special cases in which the algebraic conditions in Theorem 11 and its corollaries can be somewhat simplified; one illustrative example will suffice here.

<u>Corollary 4 to Theorem 11</u>. *If V is a normal germ of complex analytic variety having at most an isolated singularity, then the set of germs of complex analytic varieties having normalization V and also having at most an isolated singularity is in one-to-one correspondence with the set of equivalence classes of subalgebras $\mathcal{R} \subseteq {}_V\mathcal{O}$ such that $\mathcal{R}$ contains the identity and some power of the maximal ideal of ${}_V\mathcal{O}$; equivalence is as defined in Corollary 3 to Theorem 11.*

Proof. As a consequence of Corollary 3 to Theorem 11 the set of all germs of complex analytic varieties having normalization V is in one-to-one correspondence with the set of equivalence classes of subalgebras with identities $\mathfrak{R} \subseteq {}_V\mathcal{O}$ satisfying conditions (a), (b), and (c). If a subalgebra $\mathfrak{R}$ corresponds to a germ W having at most an isolated singularity, that is if $\mathfrak{R} = \varphi^*({}_W\mathcal{O})$ for some simple analytic mapping $\varphi\colon V \to W$ where W has at most an isolated singularity, and if φ is not an isomorphism, then as a consequence of Corollary 2 to Theorem 9 the conductor ${}_W\mathfrak{d}$ has the property that $\operatorname{loc} {}_W\mathfrak{d} = 0$, and the image $\varphi^*({}_W\mathfrak{d}) \subseteq {}_V\mathcal{O}$ generates an ideal $\mathfrak{A} = {}_V\mathcal{O}\cdot\varphi^*({}_W\mathfrak{d})$ in ${}_V\mathcal{O}$ such that $\operatorname{loc} \mathfrak{A} = 0$ as well; thus by the Hilbert zero theorem $\mathfrak{A} \supseteq {}_V\mathfrak{m}^N$ for some positive integer N. However it follows from the definition of universal denominator that $\mathfrak{A} = {}_V\mathcal{O}\cdot\varphi^*({}_W\mathfrak{d}) \subseteq \varphi^*({}_W\mathcal{O}) = \mathfrak{R}$; and consequently $\mathfrak{R} \supseteq \mathfrak{A} \supseteq {}_V\mathfrak{m}^N$ for some positive integer N. Of course if φ is an isomorphism then $\mathfrak{R} = {}_V\mathcal{O} \supseteq {}_V\mathfrak{m}$. On the other hand if $\mathfrak{R} \subseteq {}_V\mathcal{O}$ is a subalgebra with identity such that $\mathfrak{R} \supseteq {}_V\mathfrak{m}^N$ for some positive integer N then ${}_V\mathfrak{m} \supseteq {}_V\mathcal{O}\cdot(\mathfrak{R} \cap {}_V\mathfrak{m}) \supseteq {}_V\mathfrak{m}^N$ and $\mathfrak{R} + {}_V\mathfrak{m}^N = \mathfrak{R}$ and ${}_V\mathfrak{m}^N\cdot{}_V\mathcal{O} \subseteq \mathfrak{R}$, and consequently $\mathfrak{R}$ satisfies conditions (a), (b), and (c); there is thus a simple analytic mapping $\varphi\colon V \to W$ such that $\mathfrak{R} = \varphi^*({}_W\mathcal{O})$. Furthermore $\varphi^*({}_W\mathfrak{m}^N)\cdot{}_V\mathcal{O} \subseteq {}_V\mathfrak{m}^N \subseteq \mathfrak{R} = \varphi^*({}_W\mathcal{O})$ so that ${}_W\mathfrak{m}^N \subseteq {}_W\mathfrak{d}$, hence $\operatorname{loc} {}_W\mathfrak{d} \subseteq 0$ and W is normal outside the base point 0; thus $\varphi\colon V \to W$ is an analytic equivalence outside the base point, so that W also has at most an isolated singularity. That suffices to conclude the proof of the corollary.

(b) The preceding results can be used to approach the classification of germs of complex analytic varieties through their normalizations. All the irreducible germs of complex analytic varieties having a given normalization V are described by subalgebras of the local algebra ${}_V\mathcal{O}$, as in Theorem 11 and its corollaries. Of course this merely replaces one part of the classification problem by another problem, that of describing the equivalence classes of admissible subalgebras of the local algebras of normal germs of complex analytic varieties; but an illustration of the usefulness of this reduction will appear in the discussion of some examples later in this section of these notes. There remains the problem of classifying normal germs of complex analytic varieties; that too can be reduced to a reasonable although rather more difficult algebraic problem. First however it is convenient to establish some useful auxiliary results.

Theorem 12. If V is a normal germ of complex analytic variety and $\pi: V \rightarrow \mathbb{C}^k$ is a representation of V by a branched analytic covering, the branch points of which lie at most over a proper analytic subvariety D in $\mathbb{C}^k$, then every point of V lying over a regular point of D is necessarily a regular point of V; consequently $\dim \mathcal{S}(V) \leq \dim V - 2$.

Proof. In an open neighborhood U of any regular point of D choose a system of local coordinates $z_1,\ldots,z_k$ centered at that point such that U is a polydisc in those coordinates and $D = \{(z_1,\ldots,z_k) \in U \mid z_k = 0\}$; there is no loss of generality in the assumption that $\dim D = k-1$, since if $\dim D < k-1$ then $U - D$ is simply connected, the covering is therefore unbranched over U, and consequently the variety V is regular over U. Let V_o be a connected component of $\pi^{-1}(U)$. Recalling the Localization Lemma of CAV I, it can be assumed that $V_o \cap \pi^{-1}(0)$ consists of a single point and $\pi\colon V_o \to U$ is also a branched analytic covering, of say r sheets; and since V is normal and hence irreducible at each point, it follows from the Local Parametrization Theorem (Corollary 4 to Theorem 5 in CAV I) that the restriction $\pi\colon V_o - \pi^{-1}(D \cap U) \longrightarrow U - D \cap U$ is a connected unbranched analytic covering of r sheets. The restriction to a suitable polydisc $W \subseteq \mathbb{C}^k$ of the complex analytic mapping $\rho\colon \mathbb{C}^k \to \mathbb{C}^k$ defined by

$$\rho(t_1,\ldots,t_{k-1},t_k) = (t_1,\ldots,t_{k-1},t_k^{\,r})$$

is also an r sheeted branched analytic covering $\rho\colon W \to U$ such that the restriction $\rho\colon W - \rho^{-1}(D \cap U) \longrightarrow U - D \cap U$ is a connected unbranched analytic covering of r sheets. Since $\pi_1(U - D \cap U) \cong \mathbb{Z}$, the unbranched coverings defined by π and ρ are topologically equivalent, so there exists a topological homeomorphism $\varphi\colon V_o - \pi^{-1}(D \cap U) \longrightarrow W - \rho^{-1}(D \cap U)$ such that $\rho\varphi = \pi$; and since ρ and π locally are complex analytic homeomorphisms, the mapping φ is actually a complex analytic homeomorphism. The coordinate functions of this mapping φ are bounded analytic functions on $V_o - \pi^{-1}(D \cap U)$ which extend to analytic functions on all of V_o since V is normal; thus φ extends to a complex analytic mapping $\varphi: V_o \to \mathbb{C}^k$, and since $\rho\varphi = \pi$ for this extension by analytic continuation it follows that the extension is actually a complex analytic mapping $\varphi: V_o \to W$. Thus there results a simple analytic mapping $\varphi: V_o \to W$, which must be a complex analytic homeomorphism since W is nonsingular; and therefore V_o is nonsingular, and the proof of the theorem is thereby concluded.

Corollary 1 to Theorem 12. If V is a normal complex analytic variety and $W \subset V$ is a complex analytic subvariety such that $\dim W \leq \dim V - 2$, then any holomorphic function on $V - W$ extends to a holomorphic function on V. In particular any holomorphic function on $\mathcal{R}(V)$ extends to a holomorphic function on all of V.

Proof. The assertion is really a local one, so since V is necessarily pure dimensional then V can be represented as a branched analytic covering $\pi: V \to U$ of r sheets over an open subset $U \subseteq \mathbb{C}^k$; and the image $\pi(W) \subset U$ is a complex analytic subvariety with $\dim \pi(W) \leq k - 2$. If f is holomorphic on $V - W$ then as in Theorem 18 of CAV I there is a monic polynomial $p_f(X)$ with coefficients holomorphic on $U - \pi(W)$ such that $p_f(f) = 0$ on $V - W$. It follows as usual from the extended Riemann removable singularities theorem that the coefficients of the polynomial $p_f(X)$ extend to holomorphic functions on all of U; the coefficients and hence the roots of the polynomial are therefore locally bounded on $U - \pi(W)$, and since $p_f(f) = 0$ it follows that the values of the function f are locally bounded on $V - W$. The function f is then necessarily a weakly holomorphic function on V, and since V is normal f consequently extends to a holomorphic function on V. That proves the first assertion; and since the second assertion then follows immediately, in view of Theorem 12, the proof is thereby concluded.

To any germ f of a not identically vanishing holomorphic function at the origin in $\mathbb{C}^n$ and any germ W of complex analytic submanifold of codimension 1 at the origin in $\mathbb{C}^n$ there can be associated a non-negative integer $\nu_W(f)$ measuring the <u>order</u> of the function f along the submanifold W. To define this, choose a local coordinate system $z_1,\ldots,z_n$ centered at the origin in $\mathbb{C}^n$ and such that W is the germ of the submanifold $\{(z_1,\ldots,z_n) \in \mathbb{C}^n \mid z_n = 0\}$, consider the Taylor expansion of the function f in the form $f = \sum_{\nu=0}^{\infty} a_\nu z_n^\nu$ where $a_\nu \in {}_{n-1}\mathcal{O}$, and let $\nu_W(f)$ be the smallest integer $\nu \geq 0$ such that $a_\nu \neq 0$; it is easy to see that this is really independent of the choice of local coordinate system, since if $w_1,\ldots,w_n$ is another such local coordinate system then evidently $w_n = \sum_{\nu=1}^{\infty} b_\nu z_n^\nu$ where $b_\nu \in {}_{n-1}\mathcal{O}$ and b_1 is a unit in ${}_{n-1}\mathcal{O}$. This notion of order can be extended to meromorphic functions by setting $\nu_W(f_1/f_2) = \nu_W(f_1) - \nu_W(f_2)$, noting that this is well defined since $\nu_W(f_1f_2) = \nu_W(f_1) + \nu_W(f_2)$ whenever f_1, f_2 are holomorphic functions and are not identically zero. There results a mapping $\nu_W: {}_n\mathcal{M}^* \longrightarrow Z$, where ${}_n\mathcal{M}^*$ is the set of nonzero elements of the field ${}_n\mathcal{M}$; and it follows immediately from the definition that this mapping has the properties:

(3)
(a) $\nu_W(c) = 0$ for any nonzero complex constant c;

(b) $\nu_W(fg) = \nu_W(f) + \nu_W(g)$ for any $f,g \in {}_n\mathcal{M}^*$; and

(c) $\nu_W(f+g) \geq \min(\nu_W(f), \nu_W(g))$, with equality holding whenever $\nu_W(f) \neq \nu_W(g)$, for any $f,g \in {}_n\mathcal{M}^*$.

Note incidentally that if $h \in {}_n\mathcal{O}$ is any generator of the ideal $\text{id}\, W \subseteq {}_n\mathcal{O}$ and $f \in {}_n\mathcal{M}^*$ then $\nu_W(f)$ can be characterized as the unique integer ν such that the restriction of the function $f \cdot h^{-\nu}$ to W is a well defined, not identically vanishing meromorphic function on the submanifold W. The notion of order and this alternative characterization can be extended to some more general situations as well. If V is an irreducible germ of complex analytic variety and W is an irreducible germ of complex analytic subvariety of codimension 1 in V such that $W \not\subseteq \mathcal{S}(V)$, then $\mathcal{R}(W) \cap \mathcal{R}(V)$ is a dense open subset of a representative subvariety W; and at each point $p \in \mathcal{R}(W) \cap \mathcal{R}(V)$ the subvariety W is locally a submanifold of codimension 1 in the manifold V, hence for any function $f \in {}_V\mathcal{M}_p^*$ the order of the function f along the submanifold W is a well defined integer which will be denoted by $\nu_{W,p}(f)$. If $h \in {}_V\mathcal{O}_p$ generates the ideal $\text{id}\, W \subseteq {}_V\mathcal{O}_p$, then from the coherence of the sheaf of ideals of the subvariety W as in Theorem 7 of CAV I it follows that the function h also generates the ideal $\mathrm{id}\, W \subseteq {}_V\mathcal{O}_q$ for all points $q \in V$ sufficiently near p; and since the restriction to W of the function $f \cdot h^{-\nu}$ is a well defined, not identically vanishing meromorphic function on the submanifold W near p, it follows that $\nu_{W,p}(f) = \nu_{W,q}(f)$ for all points $q \in W$ sufficiently near p. Thus for any function $f \in {}_V\mathcal{M}^*$ the integer $\nu_{W,p}(f)$ is a locally constant function of p for all points $p \in \mathcal{R}(W) \cap \mathcal{R}(V)$ sufficiently near the base point; but since W is irreducible the set $\mathcal{R}(W) \cap \mathcal{R}(V)$ is connected, hence $\nu_{W,p}(f)$ is actually independent of the point p. This common value will be taken to be the order of the function $f \in {}_V\mathcal{M}^*$ along the subvariety W, and will be denoted by $\nu_W(f)$. It is obvious from the definition that this mapping $\nu_W\colon {}_V\mathcal{M}^* \to \mathbb{Z}$ also has the properties (3:a,b,c). It is also clear that if the ideal $\mathrm{id}\, W \subseteq {}_V\mathcal{O}$ is the principal ideal generated by a function $h \in {}_V\mathcal{O}$ and if $f \in {}_V\mathcal{M}^*$, then $\nu_W(f)$ can be characterized as the unique integer ν such that the restriction of the function $f \cdot h^{-\nu}$ to W is a well defined, not identically vanishing meromorphic function on the subvariety W. For emphasis, note again that this mapping $\nu_W\colon {}_V\mathcal{M}^* \to \mathbb{Z}$ has only been defined when $W \not\subseteq \mathcal{S}(V)$.

<u>Theorem 13</u>. If V_1, V_2 are germs of irreducible complex analytic varieties such that V_1 is normal and if $\varphi^*\colon {}_{V_2}\mathcal{M} \to {}_{V_1}\mathcal{M}$ is a homomorphism of $\mathbb{C}$-algebras with identities, then $\varphi^*({}_{V_2}\mathcal{O}) \subseteq {}_{V_1}\mathcal{O}$; consequently the homomorphism φ^* is induced by a complex analytic mapping $\varphi\colon V_1 \to V_2$.

Proof. If $\varphi^*({}_{V_2}\mathcal{O}) \subseteq {}_{V_1}\mathcal{O}$ then by Theorem 1 the restriction $\varphi^*\colon {}_{V_2}\mathcal{O} \to {}_{V_1}\mathcal{O}$ is induced by a complex analytic mapping $\varphi\colon V_1 \to V_2$, hence so is the homomorphism $\varphi^*\colon {}_{V_2}\mathcal{M} \to {}_{V_1}\mathcal{M}$; thus it is only necessary to show that $\varphi^*({}_{V_2}\mathcal{O}) \subseteq {}_{V_1}\mathcal{O}$. Suppose contrariwise that there is an element $f \in {}_{V_2}\mathcal{O}$ such that $\varphi^*(f) \notin {}_{V_1}\mathcal{O}$; the image function $\varphi^*(f) \in {}_{V_1}\mathcal{M}$ is then a meromorphic function $\varphi^*(f) = f_1/f_2$ where f_2 is a nonunit in ${}_{V_1}\mathcal{O}$. Let W_i be the irreducible components of the zero locus of the function f_2 on V_1, noting that $\dim W_i = \dim V_1 - 1$. Since V_1 is normal it follows from Theorem 12 that $\dim \mathcal{S}(V_1) \leq \dim V_1 - 2$; and therefore $W_i \not\subseteq \mathcal{S}(V_1)$ and the orders $\nu_{W_i}(f_1/f_2)$ are well defined. If $\nu_{W_i}(f_1/f_2) \geq 0$ for all the irreducible components W_i then the function f_1/f_2 is clearly holomorphic on $\mathcal{R}(V_1)$, hence from Corollary 1 to Theorem 12 it follows that $f_1/f_2 \in {}_{V_1}\mathcal{O}$ in contradiction to the assumption made above; therefore there is at least one component W_1 for which $\nu_{W_1}(f_1/f_2) < 0$. Since ${}_{V_2}\mathcal{M}$ is a field and the homomorphism $\varphi^*\colon {}_{V_2}\mathcal{M} \to {}_{V_1}\mathcal{M}$ is nontrivial the kernel of φ^* is just the zero element of ${}_{V_2}\mathcal{M}$, hence the restriction of φ^* is a homomorphism $\varphi^*\colon {}_{V_2}\mathcal{M}^* \to {}_{V_1}\mathcal{M}^*$; and the mapping $\nu\colon {}_{V_2}\mathcal{M}^* \to Z$ defined by $\nu(g) = \nu_{W_1}(\varphi^*(g))$ for any $g \in {}_{V_2}\mathcal{M}^*$ then obviously satisfies conditions (3:a,b,c). However the element $f \in {}_{V_2}\mathcal{O}$ has the property that $\nu(f) = \nu_{W_1}(f_1/f_2) < 0$; and it is easy to see that that leads to a contradiction, as follows. Choose a constant

c such that $f + c$ is a unit in ${}_{V_2}\mathcal{O}$, hence such that the function $f + c$ is nonzero near the base point of V_2; and note that for any positive integer n there is consequently a function $g \in {}_{V_2}\mathcal{O}$ such that $g^n = f + c$. From (3:b) it follows that $n \cdot \nu(g) = \nu(g^n) = \nu(f + c)$, and since $\nu(c) = 0$ by (3:a) and $\nu(f) < 0$ then as a consequence of (3:c) necessarily $\nu(f + c) = \nu(f)$; thus $n \cdot \nu(g) = \nu(f)$, hence the nonzero integer $|\nu(f)|$ is divisible by any positive integer n, which is of course impossible. That contradiction suffices to conclude the proof of the theorem.

Corollary 1 to Theorem 13. Two irreducible germs V_1, V_2 of complex analytic varieties have the same normalization if and only if their local function fields ${}_{V_1}\mathcal{M}$, ${}_{V_2}\mathcal{M}$ are isomorphic fields.

Proof. If V_1, V_2 are irreducible germs of complex analytic varieties with the respective normalizations $\hat{V}_1$, $\hat{V}_2$, then of course ${}_{\hat{V}_1}\mathcal{M} \cong {}_{V_1}\mathcal{M}$ and ${}_{\hat{V}_2}\mathcal{M} \cong {}_{V_2}\mathcal{M}$. Thus if $\hat{V}_1 = \hat{V}_2$ then the fields ${}_{V_1}\mathcal{M}$, ${}_{V_2}\mathcal{M}$ are certainly isomorphic. On the other hand any field isomorphism $\varphi^*: {}_{V_2}\mathcal{M} \to {}_{V_1}\mathcal{M}$ can be viewed as a field isomorphism $\varphi^*: {}_{\hat{V}_2}\mathcal{M} \to {}_{\hat{V}_1}\mathcal{M}$, and is also obviously an isomorphism of $\mathbb{C}$-algebras; it then follows from Theorem 13 that the isomorphism φ^* is induced by a complex analytic mapping $\varphi: \hat{V}_1 \to \hat{V}_2$, and since the inverse to φ^* is also induced by a complex analytic mapping it further follows that φ is actually an equivalence of germs of complex analytic varieties. That suffices to conclude the proof of the corollary.

The extension of this corollary to reducible germs of complex analytic varieties is quite trivial, in view of Corollary 2 to Theorem 8, so need not be gone into further. The classification of normal germs of complex analytic varieties is thus reduced to the purely algebraic problem of classifying the local function fields of irreducible germs of complex analytic varieties; when an irreducible germ V is represented by a branched analytic covering $\pi: V \rightarrow \mathbb{C}^k$ then its function field ${}_V\mathcal{M}$ is a finite algebraic extension of the local field ${}_k\mathcal{M}$ of germs of meromorphic functions at the origin in $\mathbb{C}^k$, indeed as fields ${}_V\mathcal{M} \cong {}_k\mathcal{M}[z]$ where z is algebraic over ${}_k\mathcal{M}$. Needless to say, this algebraic problem is far from trivial.

The further investigation of the local order functions $\nu_W: {}_V\mathcal{M}^* \longrightarrow \mathbb{Z}$ and their generalizations, or equivalently the study of discrete valuations of the fields ${}_V\mathcal{M}$, is another interesting topic with algebraic appeal. For work in this direction the reader is referred to Hej Iss'sa (H. Hironaka), Annals of Mathematics, Vol. 83 (1966), pages 34-46; the proof of Theorem 13 given here is based on the ideas in that paper.

(c) For one-dimensional germs of complex analytic varieties the singularities are necessarily isolated, and moreover it follows from Theorem 12 that normal germs are necessarily nonsingular. Therefore by Corollary 4 to Theorem 11 the classification of irreducible one-dimensional germs of complex analytic varieties is reduced to the classification of equivalence classes of subalgebras $\mathcal{R} \subseteq {}_1\mathcal{O}$ such that $\mathcal{R}$ contains the identity and a power of the maximal ideal of ${}_1\mathcal{O}$; indeed the classification conveniently decomposes into a limit of the relatively finite problems of classifying the equivalence classes of subalgebras $\mathcal{R} \subseteq {}_1\mathcal{O}$ such that $1 \in \mathcal{R}$ and ${}_1\mathfrak{m}^N \subseteq \mathcal{R}$ for various positive integers N. As an illustrative example this latter classification will be carried out in detail for the case $N = 5$.

Suppose first merely that $\mathcal{R} \subseteq {}_1\mathcal{O}$ is a subalgebra such that ${}_1\mathfrak{m}^5 \subseteq \mathcal{R}$; the residue class algebra $\tilde{\mathcal{R}} = \mathcal{R}/{}_1\mathfrak{m}^5$ is then a subalgebra of the five-dimensional algebra ${}_1\mathcal{O}/{}_1\mathfrak{m}^5$. An element $f \in {}_1\mathcal{O}/{}_1\mathfrak{m}^5$ can be identified with the vector in $\mathbb{C}^5$ consisting of the first five coefficients in the Taylor expansion of any representative $f \in {}_1\mathcal{O}$; addition and scalar multiplication in the algebra ${}_1\mathcal{O}/{}_1\mathfrak{m}^5$ then correspond to addition and scalar multiplication in the vector space $\mathbb{C}^5$, while multiplication has the form

$$(a_0,a_1,a_2,a_3,a_4)\cdot(b_0,b_1,b_2,b_3,b_4) = (a_0b_0,a_0b_1+a_1b_0,\dots) .$$

There are various possibilities for subalgebras $\tilde{\mathcal{R}} \subseteq {}_1\mathcal{O}/{}_1\mathfrak{m}^5$, and these can be grouped conveniently by dimension. If $\dim_{\mathbb{C}} \tilde{\mathcal{R}} = 1$ then the subalgebra $\tilde{\mathcal{R}} \subseteq {}_1\mathcal{O}/{}_1\mathfrak{m}^5$ is generated as a vector space by a single element $A = (a_0,\dots,a_4)$; and the vector subspace of ${}_1\mathcal{O}/{}_1\mathfrak{m}^5$ spanned by an element A is a subalgebra precisely when $A^2 = kA$ for some scalar $k \in \mathbb{C}$. If $a_0 \neq 0$ it can of course be assumed that $a_0 = 1$, and then

$$A^2 = (1,\ 2a_1,\ 2a_2 + a_1^2,\ 2a_3 + 2a_1a_2,\ 2a_4 + 2a_1a_3 + a_2^2)\ ;$$

and upon comparing terms it follows readily that $A^2 = kA$ if and only if $k = 1$, $a_1 = a_2 = a_3 = a_4 = 0$. If $a_0 = 0$ then

$$A^2 = (0,\ 0,\ a_1^2,\ 2a_1a_2,\ 2a_1a_3 + a_2^2)\ ;$$

and upon comparing terms it follows equally readily that $A^2 = kA$ if and only if $k = a_1 = a_2 = 0$. Thus there are only two possibilities for the generator A:

(i') $A = (1,0,0,0,0)$, in which case $A^2 = A$;

(i") $A = (0,0,0,a_3,a_4)$ for some $a_3,a_4 \in \mathbb{C}$, in which case $A^2 = 0$.

If $\dim_{\mathbb{C}} \tilde{\mathcal{R}} = 2$ then the subalgebra $\tilde{\mathcal{R}} \subseteq {}_1\mathcal{O}/{}_1\mathfrak{m}^5$ is generated as a vector space by two linearly independent vectors $A = (a_0,\ldots,a_4)$, $B = (b_0,\ldots,b_4)$; and the vector subspace of ${}_1\mathcal{O}/{}_1\mathfrak{m}^5$ spanned by two elements A, B is a subalgebra precisely when the products A^2, AB, B^2 lie in that subspace. It can always be assumed that the basis A, B is so chosen that $a_0 = \cdots = a_{\nu-1} = 0$, $a_\nu = 1$, $b_0 = \cdots = b_\nu = 0$ for some index ν with $0 \leq \nu \leq 3$; and then clearly $B^2 = kA + \ell B$ for some scalars $k,\ell \in \mathbb{C}$ only when $k = 0$ hence only when B generates a one-dimensional subalgebra of ${}_1\mathcal{O}/{}_1\mathfrak{m}^5$, in which case in view of the preceding observations necessarily $B = (0,0,0,b_3,b_4)$. If $\nu = 0$ then upon comparing terms it follows that $A^2 = kA + \ell B$ if and only if $k = 1$, $a_1 = a_2 = 0$, $a_3 = \ell b_3$, $a_4 = \ell b_4$; but then A can be replaced by $A - \ell B$, hence it can also be assumed that $a_3 = a_4 = 0$. If $\nu > 0$ then upon comparing terms it follows that $A^2 = kA + \ell B$ if and only if $a_1 = ka_2 = ka_3 + \ell b_3 = 0$, $ka_4 + \ell b_4 = a_2^2$. In these equations $k \neq 0$ implies that A, B are linearly dependent, hence necessarily $k = 0$. Next $\ell \neq 0$ implies that $b_3 = 0$, and hence it can be assumed that $b_4 = a_2 = \ell = 1$; and replacing A by $A - a_4B$ it can also be assumed that $a_4 = 0$. Finally $\ell = 0$ implies that $a_2 = 0$, and hence it can be assumed that $a_3 = 1$, $b_3 = 0$, $b_4 = 1$; and replacing A by $A - a_4B$ it can also be assumed that $a_4 = 0$. Thus there are three possibilities for the generators A, B:

(ii') $A = (1,0,0,0,0)$, $B = (0,0,0,b_3,b_4)$ for some $b_3, b_4 \in \mathbb{C}$, in which case $A^2 = A$, $AB = B$, $B^2 = 0$;

(ii") $A = (0,0,1,a_3,0)$, $B = (0,0,0,0,1)$ for some $a_3 \in \mathbb{C}$, in which case $A^2 = B$, $AB = 0$, $B^2 = 0$;

(ii'") $A = (0,0,0,1,0)$, $B = (0,0,0,0,1)$, in which case $A^2 = 0$, $AB = 0$, $B^2 = 0$.

If $\dim_{\mathbb{C}} \tilde{\mathcal{R}} = 3$ then the subalgebra $\tilde{\mathcal{R}} \subseteq {}_1\mathcal{O}/{}_1\mathfrak{m}^5$ is generated as a vector space by three linearly independent vectors A, B, C; it can be assumed that $a_0 = \dots = a_{\nu-1} = 0$, $a_\nu = 1$, $b_0 = \dots = b_\nu = c_0 = \dots = c_\nu = 0$ for some index ν with $0 \leq \nu \leq 2$, and as before the vectors B, C span a two-dimensional subalgebra of ${}_1\mathcal{O}/{}_1\mathfrak{m}^5$ which must be either of the form (ii") or of the form (ii'"). Consider first the case (ii") in which

$B = (0,0,1,b_3,0)$, $C = (0,0,0,0,1)$. If $\nu = 0$ then replacing A by $A - a_2B - a_4C$ it can be assumed that $A = (1,a_1,0,a_3,0)$; and upon comparing terms it follows that $A^2 = kA + \ell B + mC$ if and only if $k = 1$, $a_1 = a_3 = \ell = m = 0$. If $\nu > 0$ then $\nu = 1$ and it can be assumed that $A = (0,1,0,a_3,0)$; but it is easy to see that $A \cdot B$ cannot possibly lie in the subspace spanned by A, B, C, hence this case cannot occur. Consider next the case (ii''') in which $B = (0,0,0,1,0)$, $C = (0,0,0,0,1)$. If $\nu = 0$, then it can be assumed that $A = (1,a_1,a_2,0,0)$, and $A^2 = kA + \ell B + mC$ if and only if $k = 1$, $a_1 = a_2 = \ell = m = 0$. If $\nu > 0$ then it can be assumed that $A = (0,a_1,a_2,0,0)$, and $A^2 = kA + \ell B + mC$ if and only if $a_1 = k = \ell = m = 0$ and hence $a_2 = 1$. Thus there are three possibilities for the generators A, B, C:

(iii') $A = (1,0,0,0,0)$, $B = (0,0,1,b_3,0)$, $C = (0,0,0,0,1)$ for some $b_3 \in \mathbb{C}$, in which case
$A^2 = A$, $AB = B$, $AC = C$, $B^2 = C$, $BC = 0$, $C^2 = 0$;

(iii'') $A = (1,0,0,0,0)$, $B = (0,0,0,1,0)$, $C = (0,0,0,0,1)$ in which case
$A^2 = A$, $AB = B$, $AC = C$, $B^2 = 0$, $BC = 0$, $C^2 = 0$;

(iii''') $A = (0,0,1,0,0)$, $B = (0,0,0,1,0)$, $C = (0,0,0,0,1)$ in which case
$A^2 = C$, $AB = 0$, $AC = 0$, $B^2 = 0$, $BC = 0$, $C^2 = 0$.

If $\dim_{\mathbb{C}} \tilde{\mathcal{R}} = 4$ then the subalgebra $\tilde{\mathcal{R}} \subseteq {}_1\mathcal{O}/{}_1\mathcal{W}^5$ is generated as a vector space by four linearly independent vectors A, B, C, D, where it can be assumed that $a_0 = \dots = a_{\nu-1} = 0$, $a_\nu = 1$, and the first $\nu+1$ coefficients of the vectors B, C, D are all zero for some index ν with $0 \leq \nu \leq 1$. The vectors B, C, D span a three-dimensional subalgebra of ${}_1\mathcal{O}/{}_1\mathfrak{m}^5$ which must be the algebra (iii'"), and it follows easily that there are two possibilities for the generators A, B, C, D:

(iv') $A = (1,0,0,0,0)$, $B = (0,0,1,0,0)$, $C = (0,0,0,1,0)$, $D = (0,0,0,0,1)$ in which case
$A^2 = A$, $AB = B$, $AC = C$, $AD = D$, $B^2 = D$,
$BC = BD = C^2 = CD = D^2 = 0$;

(iv") $A = (0,1,0,0,0)$, $B = (0,0,1,0,0)$, $C = (0,0,0,1,0)$, $D = (0,0,0,0,1)$ in which case
$A^2 = B$, $AB = C$, $AC = D$, $B^2 = D$,
$BC = BD = C^2 = CD = D^2 = 0$.

Finally if $\dim_C \tilde{\mathcal{R}} = 5$ then

(v) $\tilde{\mathcal{R}} = {}_1\mathcal{O}/{}_1\mathfrak{m}^5$,

and the catalog of subalgebras of ${}_1\mathcal{O}/{}_1\mathfrak{m}^5$ is then complete. Of all of these only the six subalgebras (i'), (ii'), (iii'), (iii"), (iv'), (v) contain the identity element of ${}_1\mathcal{O}/{}_1\mathfrak{m}^5$; and hence the subalgebras of ${}_1\mathcal{O}$ corresponding to these are precisely the subalgebras $\mathcal{R} \subseteq {}_1\mathcal{O}$ such that $1 \in \mathcal{R}$ and ${}_1\mathfrak{m}^5 \subseteq \mathcal{R}$.

Turning next to the question of equivalences among these subalgebras, in the sense of Corollary 3 to Theorem 11, note that any automorphism of ${}_1\mathcal{O}$ preserves the ideals ${}_1\mathfrak{m}^N$ hence determines an automorphism of the residue class algebra ${}_1\mathcal{O}/{}_1\mathfrak{m}^5$.

Under these automorphisms subalgebras $\tilde{\mathcal{R}} \subseteq {}_1\mathcal{O}/{}_1\mathfrak{m}^5$ belonging to different ones of the six classes of subalgebras in the preceding catalog are never equivalent, since they are obviously not even isomorphic as algebras; therefore the only possibilities of equivalences are among the various subalgebras of class (ii') for different values of the parameters b_3, b_4 or among the various subalgebras of class (iii') for different values of the parameter b_3. Now by Theorem 1 an automorphism of ${}_1\mathcal{O}$ is induced by a nonsingular change of the local coordinate at the origin in $\mathbb{C}^1$, say of the form $z = c_1 w + c_2 w^2 + \ldots$ where $c_1 \neq 0$. For the algebras (ii') such an automorphism leaves the generator A unchanged and transforms the generator B into the vector

$B' = (0, 0, 0, c_1^3 b_3, 3c_1^2 c_2 b_3 + c_1^4 b_4)$; hence there are precisely two equivalence classes of these subalgebras, one corresponding to those algebras for which $b_3 \neq 0$ and represented by the algebra for which $B = (0,0,0,1,0)$, the other corresponding to those algebras for which $b_3 = 0$ and represented by the algebra for which $B = (0,0,0,0,1)$. For the algebras (iii') such an automorphism again leaves the generator A unchanged and transforms the generators B, C into the vectors

$B' = (0, 0, c_1^2, 2c_1 c_2 + b_3 c_1^2, 2c_1 c_3 + c_2^2 + 3c_1^2 c_2 b_3)$,

$C' = (0,0,0,0,c_1^4)$; hence all of these subalgebras are clearly equivalent, and the equivalence class can be represented by that algebra for which $b_3 = 0$. Altogether therefore there are seven equivalence classes of subalgebras $\mathcal{R} \subseteq {}_1\mathcal{O}$ such that $1 \in \mathcal{R}$

and ${}_1\mathfrak{m}^5 \subseteq \mathcal{R}$, corresponding to seven inequivalent germs of one-dimensional complex analytic varieties; and these are described by the subalgebras (i'), (ii') with $b_3 = 1, b_4 = 0$, (ii') with $b_3 = 0, b_4 = 1$, (iii') with $b_3 = 0$, (iii"), (iv'), (v).

It is perhaps of some interest to see more explicitly what the germs of varieties are that have just been described so algebraically. In the case (i') note that $\mathcal{R} = \mathbb{C} + {}_1\mathfrak{m}^5 \subseteq {}_1\mathcal{O}$, hence the maximal ideal of the algebra $\mathcal{R}$ is ${}_{\mathcal{R}}\mathfrak{m} = {}_1\mathfrak{m} \cap \mathcal{R} = {}_1\mathfrak{m}^5$ and ${}_{\mathcal{R}}\mathfrak{m}^2 = {}_1\mathfrak{m}^{10}$; therefore $\dim_{\mathbb{C}} {}_{\mathcal{R}}\mathfrak{m} / {}_{\mathcal{R}}\mathfrak{m}^5 = 5$, and indeed the functions z^5, z^6, z^7, z^8, z^9 in ${}_{\mathcal{R}}\mathfrak{m}$ represent a basis for the complex vector space ${}_{\mathcal{R}}\mathfrak{m} / {}_{\mathcal{R}}\mathfrak{m}^2$. It then follows from Corlllary 1 to Theorem 11 that the germ at the origin of the analytic mapping $\varphi\colon \mathbb{C}^1 \to \mathbb{C}^5$ defined by $\varphi(z) = (z^5, z^6, z^7, z^8, z^9)$ has as its image the germ of a complex analytic subvariety V at the origin in $\mathbb{C}^5$ such that ${}_V\mathcal{O} \cong \varphi^*({}_V\mathcal{O}) = \mathcal{R} \subseteq {}_1\mathcal{O}$; moreover the imbedding dimension of V is 5, so that V is neatly imbedded in $\mathbb{C}^5$ and the germ of variety it represents cannot also be represented by the germ of a complex analytic subvariety in $\mathbb{C}^n$ for any $n < 5$. Note that the natural projection from $\mathbb{C}^5$ to the first coordinate axis exhibits the subvariety V as a five-sheeted branched analytic covering of $\mathbb{C}^1$, and that the second coordinate in $\mathbb{C}^5$ separates the sheets of this covering; therefore the given coordinates in $\mathbb{C}^5$ are a strictly regular system of coordinates for the ideal $\mathrm{id}\, V \subseteq {}_5\mathcal{O}$, and the canonical equations for this ideal can be deduced quite easily from the parametric representation of V given by the mapping φ.

Letting (z_1,z_2,z_3,z_4,z_5) be the given coordinates in the ambient space $\mathbb{C}^5$, the first set of canonical equations for the ideal of V are

$$p_2(z_1;z_2) = z_2^5 - z_1^6 \quad , \quad p_3(z_1;z_3) = z_3^5 - z_1^7 ,$$

$$p_4(z_1;z_4) = z_4^5 - z_1^8 \quad , \quad p_5(z_1;z_5) = z_5^5 - z_1^9 ;$$

the discriminant of the polynomial $p_2 \in {}_1\mathcal{O}[z_2]$ is $d = z_1^{24} \in {}_1\mathcal{O}$, except for a constant factor which is irrelevant here, and the second set of canonical equations for the ideal of V are

$$q_3(z_1;z_2,z_3) = z_1^{24}z_3 - z_1^{23}z_2^2 \quad , \quad q_4(z_1;z_2,z_4) = z_1^{24}z_4 - z_1^{22}z_2^3 ,$$

$$q_5(z_1;z_2,z_5) = z_1^{24}z_5 - z_1^{21}z_2^4 .$$

The latter equations can of course be simplified by dividing each by a suitable power of z_1, since $z_1 \notin \text{id } V$ and $\text{id } V$ is a prime ideal. As usual the subvariety V, outside the critical locus $z_1 = 0$ of the branched analytic covering induced by the natural projection $\mathbb{C}^5 \to \mathbb{C}^1$, is described precisely by the equations $p_2 = q_3 = q_4 = q_5 = 0$; but the complete subvariety of $\mathbb{C}^5$ described by these equations is clearly $V \cup L$ where L is the three-dimensional linear subspace defined by the equations $z_1 = z_2 = 0$. However all the canonical equations together in this case do describe precisely the subvariety V, so that

$$V = \{z \in \mathbb{C}^5 \mid p_2(z) = p_3(z) = p_4(z) = p_5(z) = q_3(z) = q_4(z) = q_5(z) = 0\}.$$

In the case (ii') with $b_3 = 1, b_4 = 0$ note that $\mathcal{R} \subseteq {}_1\mathcal{O}$ is the subalgebra consisting of the power series $\sum_{\nu=0}^{\infty} c_\nu z^\nu \in {}_1\mathcal{O}$ for which $c_1 = c_2 = c_4 = 0$; hence $\dim_{\mathbb{C}} {}_{\mathcal{R}}\mathfrak{m} / {}_{\mathcal{R}}\mathfrak{m}^2 = 3$, and the functions z^3, z^5, z^7 in ${}_{\mathcal{R}}\mathfrak{m}$ represent a basis for the complex vector space ${}_{\mathcal{R}}\mathfrak{m} / {}_{\mathcal{R}}\mathfrak{m}^2$. By Corollary 1 to Theorem 11 the subalgebra $\mathcal{R}$ then corresponds to the germ at the origin in $\mathbb{C}^3$ of the complex analytic subvariety V described parametrically by the mapping $\varphi\colon \mathbb{C}^1 \to \mathbb{C}^3$ for which $\varphi(z) = (z^3, z^5, z^7)$. The given coordinates (z_1, z_2, z_3) in $\mathbb{C}^3$ are again a strictly regular system of coordinates for the ideal $\operatorname{id} V \subseteq {}_3\mathcal{O}$, and the canonical equations for the ideal of V are

$$p_2(z_1; z_2) = z_2^3 - z_1^5 \quad , \quad p_3(z_1; z_3) = z_3^3 - z_1^7$$

$$q_3(z_1; z_2, z_3) = z_1^{10} z_3 - z_1^9 z_2^2$$

and

$$V = \{ z \in \mathbb{C}^5 \mid p_2(z) = p_3(z) = q_3(z) = 0 \} .$$

In the case (ii') with $b_3 = 0, b_4 = 1$ note that $\mathcal{R} \subseteq {}_1\mathcal{O}$ is the subalgebra $\mathbb{C} + {}_1\mathfrak{m}^4 \subseteq {}_1\mathcal{O}$, hence $\dim {}_{\mathcal{R}}\mathfrak{m} / {}_{\mathcal{R}}\mathfrak{m}^2 = 4$ and the functions z^4, z^5, z^6, z^7 in ${}_{\mathcal{R}}\mathfrak{m}$ represent a basis for the complex vector space ${}_{\mathcal{R}}\mathfrak{m} / {}_{\mathcal{R}}\mathfrak{m}^2$; the subalgebra $\mathcal{R}$ then corresponds to the germ at the origin in $\mathbb{C}^4$ of the complex analytic subvariety V described parametrically by the mapping $\varphi\colon \mathbb{C}^1 \to \mathbb{C}^4$ for which $\varphi(z) = (z^4, z^5, z^6, z^7)$. The coordinates in $\mathbb{C}^4$ are a strictly regular system of coordinates for the ideal $\operatorname{id} V \subseteq {}_4\mathcal{O}$, and the canonical equations for the ideal of V are

$$p_2(z_1;z_2) = z_2^4 - z_1^5 \quad , \quad p_3(z_1;z_3) = z_3^2 - z_1^3 \quad , \quad p_4(z_1,z_4) = z_4^4 - z_1^7 ,$$

$$q_3(z_1;z_2,z_3) = z_1^{15}z_3 - z_1^{14}z_2^2 \quad , \quad q_4(z_1;z_2,z_4) = z_1^{15}z_4 - z_1^{13}z_2^3 ;$$

and

$$V = \{z \in \mathbb{C}^4 \mid p_2(z) = p_3(z) = p_4(z) = q_3(z) = q_4(z) = 0\}$$

In the case (iii') with $b_3 = 0$ note that $\mathcal{R} \subseteq {}_1\mathcal{O}$ is the subalgebra consisting of the power series $\sum_{\nu=0}^{\infty} c_\nu z^\nu \in {}_1\mathcal{O}$ for which $c_1 = c_3 = 0$, hence $\dim_{\mathbb{C}}\, {}_{\mathcal{R}}\mathfrak{m} / {}_{\mathcal{R}}\mathfrak{m}^2 = 2$ and the functions z^2, z^5 in ${}_{\mathcal{R}}\mathfrak{m}$ represent a basis for the complex vector space ${}_{\mathcal{R}}\mathfrak{m} / {}_{\mathcal{R}}\mathfrak{m}^2$; the subalgebra $\mathcal{R}$ then corresponds to the germ at the origin in $\mathbb{C}^2$ of the complex analytic subvariety V described parametrically by the mapping $\varphi\colon \mathbb{C}^1 \to \mathbb{C}^2$ for which $\varphi(z) = (z^2,z^5)$. The canonical equation for the ideal $\operatorname{id} V \subseteq {}_2\mathcal{O}$ is

$$p_2(z_1,z_2) = z_2^2 - z_1^5 ,$$

and V is the hypersurface

$$V = \{z \in \mathbb{C}^2 \mid p_2(z) = 0\} .$$

In the case (iii") note that $\mathcal{R} \subseteq {}_1\mathcal{O}$ is the subalgebra $\mathcal{R} = \mathbb{C} + {}_1\mathfrak{m}^3 \subseteq {}_1\mathcal{O}$, hence $\dim_{\mathbb{C}}\, {}_{\mathcal{R}}\mathfrak{m} / {}_{\mathcal{R}}\mathfrak{m}^2 = 3$ and the functions z^3, z^4, z^5 in ${}_{\mathcal{R}}\mathfrak{m}$ represent a basis for the complex vector space

${}_R\mathfrak{m}/{}_R\mathfrak{m}^2$; the subalgebra $\mathcal{R}$ then corresponds to the germ at the origin in $\mathbb{C}^3$ of the complex analytic subvariety V described parametrically by the mapping $\varphi\colon \mathbb{C}^1 \to \mathbb{C}^3$ for which $\varphi(z) = (z^3, z^4, z^5)$. The coordinates in $\mathbb{C}^3$ are a strictly regular system of coordinates for the ideal $\text{id}\, V \subseteq {}_3\mathcal{O}$, and the canonical equations for the ideal of V are

$$p_2(z_1;z_2) = z_2^3 - z_1^4 \quad , \quad p_3(z_1;z_3) = z_3^3 - z_1^5 \, ,$$

$$q_3(z_1;z_2,z_3) = z_1^8\ z_3 - z_1^7 z_2^2 \, ,$$

and

$$V = \{z \in \mathbb{C}^3 \mid p_2(z) = p_3(z) = q_3(z) = 0\} \, .$$

In case (iv') note that $\mathcal{R} \subseteq {}_1\mathcal{O}$ is the subalgebra $\mathcal{R} = \mathbb{C} + {}_1\mathfrak{m}^2$, hence $\dim {}_R\mathfrak{m}/{}_R\mathfrak{m}^2 = 2$ and the functions z^2, z^3 in ${}_R\mathfrak{m}$ represent a basis for the complex vector ${}_R\mathfrak{m}/{}_R\mathfrak{m}^2$; the subalgebra $\mathcal{R}$ then corresponds to the germ at the origin in $\mathbb{C}^2$ of the complex analytic subvariety V described parametrically by the mapping $\varphi\colon \mathbb{C}^1 \to \mathbb{C}^2$ for which $\varphi(z) = (z^2, z^3)$. The canonical equation for the ideal $\text{id}\, V \subseteq {}_2\mathcal{O}$ is

$$p_2(z_1;z_2) = z_2^2 - z_1^3$$

and V is the hypersurface

$$V = \{z \in \mathbb{C}^2 \mid p_2(z) = 0\} \, .$$

In the case (v) of course $\mathcal{R} = {}_1\mathcal{O}$, and the subalgebra $\mathcal{R}$

corresponds to the germ of a regular analytic variety. These observations are summarized in Table 1.

A few further comments about these examples should also be inserted here. It is apparent upon examining Table 1 that the characteristic ideal of the mapping φ does not determine that mapping fully; but in this special case the characteristic ideal does have an interesting interpretation as suggested by that table, namely, the characteristic ideal is of the form $\mathfrak{A} = {}_1\mathfrak{m}^r$ where r is the smallest integer such that the germ V can be represented by a branched analytic covering $V \to C^1$ of r sheets. The proof is quite straightforward and will be left as an exercise to the reader. Although some readers may feel that this exercise in classification has already been carried too far, it has nonetheless not been carried out far enough to illustrate one important phenomenon. In the classification of the subalgebras $\mathfrak{R} \subseteq {}_1\mathcal{O}$ such that $1 \in \mathfrak{R}$ and ${}_1\mathfrak{m}^N \subseteq \mathfrak{R}$ for $N = 5$ there appeared some families of subalgebras depending on auxiliary parameters; for example the family of subalgebras (ii') depends on the parameters b_3, b_4, which can be arbitrary complex numbers not both of which are zero. These parameters disappeared when passing to equivalence classes of subalgebras; for example in the family of subalgebras (ii') the equivalence class was determined merely by whether the parameter b_3 is zero, hence there were just two equivalence classes. However for larger values of N the equivalence classes of subalgebras of ${}_1\mathcal{O}$ and hence the germs of complex analytic varieties they describe will generally depend on some auxiliary

Table 1

Germs of one-dimensional irreducible complex analytic varieties V with normalization $\varphi: \mathbb{C}^1 \to V$ such that ${}_1\mathfrak{m}^5 \subseteq \varphi^*({}_V\mathcal{O}) \subseteq {}_1\mathcal{O}$. (Column 1: defining equations for V; column 2: parametrization by the normalization φ; column 3: local ring $\mathfrak{R} = \varphi^*({}_V\mathcal{O}) \subseteq {}_1\mathcal{O}$ of V; column 4: characteristic ideal $\mathfrak{A} = {}_V\mathcal{O}\cdot\varphi^*({}_V\mathfrak{m})$ of φ; column 5: imbedding dimension of V; column 6: reference to the preceding discussion.)

1: $V =$	2: $\varphi(z) =$	3: $\mathfrak{R} =$	4: $\mathfrak{A} =$	5	6
regular analytic variety	(z)	${}_1\mathcal{O}$	${}_1\mathfrak{m}$	1	(v)
$z_2^2 = z_1^3$	(z^2, z^3)	$\mathbb{C} + {}_1\mathfrak{m}^2$	${}_1\mathfrak{m}^2$	2	(iv')
$z_2^2 = z_1^5$	(z^2, z^5)	$\mathbb{C} + \mathbb{C}z^2 + {}_1\mathfrak{m}^4$	${}_1\mathfrak{m}^2$	2	(iii')
$z_2^3 = z_1^4$, $z_3^3 = z_1^5$, $z_1z_3 = z_2^2$	(z^3, z^4, z^5)	$\mathbb{C} + {}_1\mathfrak{m}^3$	${}_1\mathfrak{m}^3$	3	(iii")
$z_2^3 = z_1^5$, $z_3^3 = z_1^7$, $z_1z_3 = z_2^2$	(z^3, z^5, z^7)	$\mathbb{C} + \mathbb{C}z^3 + {}_1\mathfrak{m}^5$	${}_1\mathfrak{m}^3$	3	(ii')
$z_2^4 = z_1^5$, $z_3^2 = z_1^3$, $z_4^4 = z_1^7$, $z_1z_3 = z_2^2$, $z_1^2z_4 = z_2^3$	(z^4, z^5, z^6, z^7)	$\mathbb{C} + {}_1\mathfrak{m}^4$	${}_1\mathfrak{m}^4$	4	(ii')
$z_2^5 = z_1^6$, $z_3^5 = z_1^7$, $z_4^5 = z_1^8$, $z_5^5 = z_1^9$, $z_1z_3 = z_2^2$, $z_1^2z_4 = z_2^3$, $z_1^3z_5 = z_2^4$	$(z^5, z^6, z^7, z^8, z^9)$	$\mathbb{C} + {}_1\mathfrak{m}^5$	${}_1\mathfrak{m}^5$	5	(i')

parameters. For example consider the class of subalgebras $\mathcal{R} \subseteq {}_1\mathcal{O}$ of the form

$$\mathcal{R} = \mathbb{C} + \mathbb{C}z^6 + \mathbb{C}(z^9 + a_{10}z^{10} + a_{11}z^{11}) + {}_1\mathfrak{m}^{12}$$

for arbitrary complex constants a_{10}, a_{11}. Introducing a change of variable of the form $z = c_1w + c_2w^2 + \ldots$ where $c_1 \neq 0$, it is easy to see that the resulting automorphism of ${}_1\mathcal{O}$ transforms $\mathcal{R}$ into a subalgebra of precisely the same form if and only if $c_2 = c_3 = c_5 - c_1c_4a_{10} = c_6 - c_1^2 c_4a_{11} = 0$, and that then

$$\mathcal{R} = C + Cw^6 + C(w^9 + c_1a_{10}w^{10} + c_1^2 a_{11}w^{11}) + {}_1\mathfrak{m}^{12} .$$

Therefore two subalgebras of ${}_1\mathcal{O}$ of this form, corresponding to parameters (a_{10}, a_{11}) and (a'_{10}, a'_{11}) for which $a_{10}a'_{10} \neq 0$, are equivalent if and only if $a_{11}a_{10}^{-2} = (a'_{11})(a'_{10})^{-2}$; consequently the set of equivalence classes of subalgebras $\mathcal{R} \subseteq {}_1\mathcal{O}$ of this form for which $a_{10} \neq 0$ is in one-to-one correspondence with the set $\mathbb{C}$ of all complex numbers under the correspondence which associates to such a subalgebra the parameter $a = a_{11}a_{10}^{-2}$.

The goal here has merely been to discuss systematically some illustrative examples, so no attempt will be made at present to treat the classification of one-dimensional germs of complex analytic varieties in general or to examine in greater detail further properties of this special case. There is an extensive literature devoted to the study of one-dimensional germs of complex analytic varieties, especially those of imbedding dimension two (singularities of plane curves); for that the reader is referred to the following books and to the further references listed therein: R. J. Walker, Algebraic Curves, (Princeton University Press, 1950); J. G. Semple and G. T. Kneebone, Algebraic Curves (Oxford University Press, 1959); O. Zariski, Algebraic Surfaces (second edition, Springer-Verlag, 1971). A recent survey with current references is by Lê Dũng Tráng, Noeds Algébriques, Ann. Inst. Fourier, Grenoble, vol. 23 (1972), pp. 117-126.

(d) The classification of germs of two-dimensional irreducible complex analytic varieties having at most isolated singularities and having regular normalizations can also be reduced to a sequence of simple and relatively finite purely algebraic problems by applying Corollary 4 to Theorem 11; and although the treatment is, except for further complications in the details, almost an exact parallel to that of germs of one-dimensional irreducible complex analytic varieties, it is perhaps worth carrying out in some simple cases just in order to furnish a few explicit examples of higher-dimensional singularities. Consider then the problem of determining all the germs of two-dimensional complex analytic varieties V with a normalization $\varphi\colon \mathbb{C}^2 \to V$ such that ${}_2\mathfrak{m}^3 \subseteq \varphi^*({}_V\mathcal{O})$, or equivalently, the problem of determining the equivalence classes of subalgebras $\mathcal{R} \subseteq {}_2\mathcal{O}$ for which $1 \in \mathcal{R}$ and ${}_2\mathfrak{m}^3 \subseteq \mathcal{R}$.

If $\mathcal{R} \subseteq {}_2\mathcal{O}$ is any subalgebra such that ${}_2\mathfrak{m}^3 \subseteq \mathcal{R}$ then the residue class algebra $\tilde{\mathcal{R}} = \mathcal{R}/{}_2\mathfrak{m}^3$ is a subalgebra of the six-dimensional algebra ${}_2\mathcal{O}/{}_2\mathfrak{m}^3$; an element $f \in {}_2\mathcal{O}/{}_2\mathfrak{m}^3$ can be identified with the vector

$$(c_{00}, c_{10}, c_{01}, c_{20}, c_{11}, c_{02}) \in \mathbb{C}^6$$

consisting of the coefficients of the terms of at most second order in the Taylor expansion of any representative function $f \in {}_2\mathcal{O}$, and $\tilde{\mathcal{R}}$ can then be described by the vectors of a basis for the vector subspace $\tilde{\mathcal{R}} \subseteq \mathbb{C}^6$. It is a straightforward matter to list all the possibilities, just as in the case of one-dimensional varieties; but the procedure can be simplified further, since only equivalence classes of subalgebras of ${}_2\mathcal{O}/{}_2\mathfrak{m}^3$ are really of interest. A nonsingular linear change of coordinates in $\mathbb{C}^2$ induces an equivalent nonsingular linear transformation of the two-dimensional space of the coordinates (c_{10}, c_{01}) of the algebra ${}_2\mathcal{O}/{}_2\mathfrak{m}^3$; hence it can be assumed that the projection of the subalgebra $\tilde{\mathcal{R}} \subseteq {}_2\mathcal{O}/{}_2\mathfrak{m}^3$ to the two-dimensional space of the coordinates (c_{10}, c_{01}) is either 0, or the vector subspace spanned by the vector $(1,0)$, or the entire two-dimensional vector space. After this preliminary simplification it is easy to see that there are just eight classes of subalgebras $\tilde{\mathcal{R}} \subseteq {}_2\mathcal{O}/{}_2\mathfrak{m}^3$ with $1 \in \tilde{\mathcal{R}}$, with the following generators and algebra structure:

(i) $A = (1,0,0,0,0,0)$; $A^2 = A$;

(ii) $A = (1,0,0,0,0,0)$, $B = (0,0,0,b_{20},b_{11},b_{02})$; $A = 1$, $B^2 = 0$;

(iii') $A = (1,0,0,0,0,0)$, $B = (0,1,0,0,b_{11},b_{02})$, $C = (0,0,0,1,0,0)$; $A = 1$, $B^2 = C$, $BC = C^2 = 0$

(iii") $A = (1,0,0,0,0,0)$, $B = (0,0,0,b_{20},b_{11},b_{02})$,

$C = (0,0,0,c_{20},c_{11},c_{02})$; $A = 1$, $B^2 = BC = C^2 = 0$;

(iv') $A = (1,0,0,0,0,0)$, $B = (0,1,0,0,b_{11},b_{02})$,

$C = (0,0,0,1,0,0)$, $D = (0,0,0,0,d_{11},d_{02})$;

$A = 1$, $B^2 = C$, $BC = BD = C^2 = CD = D^2 = 0$;

(iv") $A = (1,0,0,0,0,0)$, $B = (0,0,0,1,0,0)$,

$C = (0,0,0,0,1,0)$, $D = (0,0,0,0,0,1)$;

$A = 1$, $B^2 = BC = BD = C^2 = CD = D^2 = 0$;

(v) $A = (1,0,0,0,0,0)$, $B = (0,1,0,0,0,0)$,

$C = (0,0,0,1,0,0)$, $D = (0,0,0,0,1,0)$,

$E = (0,0,0,0,0,1)$; $A = 1$, $B^2 = C$,

$BC = BD = BE = C^2 = CD = CE = D^2 = DE = E^2 = 0$;

(vi) $A = (1,0,0,0,0,0)$, $B = (0,1,0,0,0,0)$,

$C = (0,0,1,0,0,0)$, $D = (0,0,0,1,0,0)$,

$E = (0,0,0,0,1,0)$, $F = (0,0,0,0,0,1)$;

$A = 1$, $B^2 = D$, $BC = E$, $BD = BE = BF = 0$,

$C^2 = F$, $CD = CE = CF = D^2 = DE = DF = E^2 = EF = F^2 = 0$.

As in the discussion of one-dimensional varieties these classes are indexed by the dimension of the complex vector space $\tilde{\mathcal{R}}$. The details of the verification shed no further light and consequently will be omitted.

It is clear that further equivalences can only occur among subalgebras belonging to the same class; hence it only remains to determine which parameter values lead to equivalent subalgebras in classes (ii), (iii'), (iii"), and (iv'). In classes (ii) and (iii") the preliminary simplification is unnecessary, since the projection of the subalgebra $\tilde{\mathcal{R}} \subseteq {}_2\mathcal{O}/{}_2\mathfrak{m}^3$ to the two-dimensional space of the coordinates (c_{10}, c_{01}) is necessarily 0; hence equivalences arise from arbitrary automorphisms of ${}_2\mathcal{O}$. The effect of an automorphism of ${}_2\mathcal{O}$ on the vector B in a subalgebra of class (ii) is evidently just that of a nonsingular linear change of variables on the quadratic form $b_{20}Z_1^2 + b_{11}Z_1Z_2 + b_{02}Z_2^2$; hence the vector B can be reduced to one of the normal forms $B = (0,0,0,1,0,0)$ or $B = (0,0,0,1,0,1)$ depending on the rank of that quadratic form. The situation is almost the same for a subalgebra of class (iii"), except that then it is a matter of reducing to normal form a two-dimensional linear family of quadratic forms; and depending on whether that family contains only one or more than one singular quadratic form the vectors B, C can be reduced to one of the normal forms $B = (0,0,0,1,0,0)$, $C = (0,0,0,0,0,1)$ or $B = (0,0,0,1,0,0)$, $C = (0,0,0,0,1,0)$. In classes (iii') and (iv') the preliminary simplification is invoked to reduce the projection of the subalgebra $\tilde{\mathcal{R}} \subseteq {}_2\mathcal{O}/{}_2\mathfrak{m}^3$ to the two-dimensional space of the coordinates (c_{10}, c_{01}) to the linear subspace spanned by the vector (1,0); hence further equivalences can only arise from automorphisms of ${}_2\mathcal{O}$ the linear parts of which preserve that subspace. An automorphism of ${}_2\mathcal{O}$ with linear part the identity can be used to reduce the vector B in a subalgebra of class (iii') to the normal form $B = (0,1,0,0,0,0)$. For an algebra of class (iv') the quadratic part of an automorphism of

${}_2\mathcal{O}$ can be used to reduce the vector B to the normal form $B = (0,1,0,0,0,0)$; and an admissible linear part of an automorphism of ${}_2\mathcal{O}$ can be used to reduce the vector D to one of the normal forms $D = (0,0,0,0,1,0)$ in $D = (0,0,0,0,0,1)$. Altogether then there are eleven equivalence classes of subalgebras $\mathcal{R} \subseteq {}_2\mathcal{O}$ such that $1 \in \mathcal{R}$ and ${}_2\mathfrak{m}^3 \subseteq \mathcal{R}$, corresponding to eleven inequivalent germs of two-dimensional complex analytic varieties; and these are represented by the subalgebras (i), (ii) with $b_{20} = 1$, $b_{11} = b_{02} = 0$, (ii) with $b_{20} = b_{02} = 1$, $b_{11} = 0$, (iii') with $b_{11} = b_{02} = 0$, (iii") with $b_{20} = c_{02} = 1$, $b_{11} = b_{02} = c_{20} = c_{11} = 0$, (iii") with $b_{20} = c_{11} = 1$, $b_{11} = b_{02} = c_{20} = c_{02} = 0$, (iv') with $d_{11} = 1$, $d_{02} = b_{11} = b_{02} = 0$, (iv') with $d_{02} = 1$, $d_{11} = b_{11} = b_{02} = 0$, (iv"), (v), and (vi).

It is again of some interest to see more explicitly what the germs of varieties described by these subalgebras really are, but only the cases of relatively low imbedding dimension will be discussed in much detail to avoid what are actually rather dull complications. In case (vi) the subalgebra is $\mathcal{R} = {}_2\mathcal{O}$; that corresponds to a regular two-dimensional variety, about which nothing more needs to be said. In case (v) the subalgebra is $\mathcal{R} = \mathbb{C} + \mathbb{C}t_1 + {}_2\mathfrak{m}^2 \subseteq {}_2\mathcal{O}$, where (t_1,t_2) are local coordinates at the origin in the normalization $\mathbb{C}^2$; and it is easy to see that $\dim_{\mathbb{C}} {}_{\mathcal{R}}\mathfrak{m} / {}_{\mathcal{R}}\mathfrak{m}^2 = 4$, indeed that the functions t_1, t_2^2, t_2^3, t_1t_2 represent a basis for the complex vector space ${}_{\mathcal{R}}\mathfrak{m} / {}_{\mathcal{R}}\mathfrak{m}^2$. It then follows from Corollary 1 to Theorem 11 that the germ at the origin of the analytic mapping $\varphi: \mathbb{C}^2 \to \mathbb{C}^4$ defined by

$\varphi(t_1,t_2) = (t_1,t_2^2,t_2^3,t_1t_2)$ has as its image the germ of a complex analytic subvariety V at the origin in $\mathbb{C}^4$ such that ${}_V\mathcal{O} \cong \varphi^*({}_V\mathcal{O}) = \mathcal{R} \subseteq {}_2\mathcal{O}$. Note that the natural projection from $\mathbb{C}^4$ to the subspace $\mathbb{C}^2 \subset \mathbb{C}^4$ spanned by the first two coordinate axes exhibits the subvariety V as a two-sheeted branched analytic covering of $\mathbb{C}^2$, and that the third coordinate in $\mathbb{C}^4$ separates the sheets of this covering; therefore the coordinates in $\mathbb{C}^4$ are a strictly regular system of coordinates for the ideal $\text{id}\, V \subseteq {}_4\mathcal{O}$, and the canonical equations for that ideal can be deduced quite easily from the parametric representation of V given by the mapping φ. Letting (z_1,z_2,z_3,z_4) be the natural coordinates in $\mathbb{C}^4$, the first set of canonical equations for the ideal of V are

$$p_3(z_1,z_2;z_3) = z_3^2 - z_2^3 \quad , \quad p_4(z_1,z_2;z_4) = z_4^2 - z_1^2 z_2 \ ;$$

the discriminant of the polynomial $p_3 \in {}_2\mathcal{O}[z_3]$ is $d = z_2^3 \in {}_2\mathcal{O}$, and the second set of canonical equations for the ideal of V consists of the single equation

$$q_4(z_1,z_2;z_3,z_4) = z_2^3 z_4 - z_1 z_2^2 z_3 = z_2^2(z_2z_4 - z_1z_3) \ .$$

The ideal of V is prime, and it does not contain the function z_2 since V is neatly imbedded in $\mathbb{C}^4$; and consequently the factor $z_2z_4 - z_1z_3$ of q_4 also belongs to the ideal of V. The branch locus of the projection from V to $\mathbb{C}^2$ is as usual contained in the discriminant locus

$$D = \{z = (z_1,z_2,z_3,z_4) \in \mathbb{C}^4 \mid z_2 = 0\} ,$$

and outside D the subvariety V is described in terms of the canonical equations by $p_3 = q_4 = 0$ as a consequence of the local parametrization theorem; and if $z_2 = 0$ then $p_3(z) = q_4(z) = 0$ precisely when $z_3 = 0$ as well, so that
$\{z \in \mathbb{C}^4 \mid p_3(z) = q_4(z) = 0\} = V \cup L$ where L is the two-dimensional linear subvariety defined by the equations $z_2 = z_3 = 0$. It is clear though that all the canonical equations together define precisely the subvariety V, so that

$$V = \{z \in \mathbb{C}^4 \mid p_3(z) = p_4(z) = q_4(z) = 0\} .$$

In case (iv') with $d_{02} = 1$, $d_{11} = b_{11} = b_{02} = 0$ note that $\mathcal{R} \subseteq {}_2\mathcal{O}$ is the subalgebra $\mathcal{R} = \mathbb{C} + \mathbb{C}t_1 + \mathbb{C}t_1^2 + \mathbb{C}t_2^2 + {}_2\mathfrak{m}^3$; hence $\dim {}_{\mathcal{R}}\mathfrak{m}/{}_{\mathcal{R}}\mathfrak{m}^2 = 4$, and the functions t_1, t_2^2, t_2^3, $t_1^2t_2$ in ${}_{\mathcal{R}}\mathfrak{m}$ represent a basis for ${}_{\mathcal{R}}\mathfrak{m}/{}_{\mathcal{R}}\mathfrak{m}^2$. The subvariety V corresponding to this subalgebra is described parametrically by the mapping $\varphi\colon \mathbb{C}^2 \to \mathbb{C}^4$ defined by $\varphi(t_1,t_2) = (t_1,t_2^2,t_2^3,t_1^2t_2)$; the coordinates in $\mathbb{C}^4$ are a strictly regular system of coordinates for the ideal of V, and the canonical equations are

$$p_3(z_1,z_2;z_3) = z_3^2 - z_2^3 \qquad \text{with discriminant } d = z_2^3,$$

$$p_4(z_1,z_2;z_4) = z_4^2 - z_1^4z_2 \quad , \quad q_4(z_1,z_2;z_3,z_4) = z_2^3z_4 - z_1^2z_2^2z_3 .$$

All of these equations together again determine precisely the subvariety V. In case (iv') with $d_{11} = 1$, $d_{02} = b_{11} = b_{02} = 0$, note that $\mathcal{R} = \mathbb{C} + \mathbb{C}t_1 + \mathbb{C}t_1^2 + \mathbb{C}t_1t_2 + {}_2\mathfrak{m}^3$ and $\dim {}_{\mathcal{R}}\mathfrak{m}/{}_{\mathcal{R}}\mathfrak{m}^2 = 6$; the subvariety corresponding to this subalgebra is described parametrically by the mapping $\varphi\colon \mathbb{C}^2 \to \mathbb{C}^6$ defined by $\varphi(t_1,t_2) = (t_1,t_2^3,t_1t_2,t_1t_2^2,t_2^4,t_2^5)$, the coordinates in $\mathbb{C}^6$ are a strictly regular system of coordinates for the ideal of V, and the canonical equations are

$$p_3(z_1,z_2;z_3) = z_3^3 - z_1^3z_2 \qquad \text{with discriminant} \quad d = z_1^6z_2^2\ ,$$

$$p_4(z_1,z_2;z_4) = z_4^3 - z_1^3z_2^2\ , \quad q_4(z_1,z_2;z_3,z_4) = z_1^6z_2^2z_4 - z_1^5z_2^2z_3^2\ ,$$

$$p_5(z_1,z_2;z_5) = z_5^3 - z_2^4\ , \quad q_5(z_1,z_2;z_3,z_5) = z_1^6z_2^2z_5 - z_1^5z_2^3z_3\ ,$$

$$p_6(z_1,z_2;z_6) = z_6^3 - z_2^5\ , \quad q_6(z_1,z_2;z_3,z_6) = z_1^6z_2^2z_6 - z_1^4z_2^3z_3^2\ .$$

In this case the canonical equations do not suffice to describe the subvariety V precisely, although of course as a consequence of the local parametrization theorem they do describe the subvariety V outside the discriminant locus $D = \{z \in \mathbb{C}^6 \mid z_1z_2 = 0\}$ and V is a subvariety of the set of common zeros of the canonical equations. However note that if $z \in V$ and $z_2 = 0$ then $z = \varphi(t,0) = (t,0,0,0,0,0)$ for some parameter value t, while if $z \in V$ and $z_1 = 0$ then $z = \varphi(0,t) = (0,t^3,0,0,t^4,t^5)$ for some parameter value t; but on the other hand if z is a point at which all the canonical equations vanish, indeed even at which the nontrivial factors of the second set of canonical equations also vanish, then if $z_2 = 0$ clearly $z_3 = z_4 = z_5 = z_6 = 0$ so that $z \in V$, while if $z_1 = 0$ then $z_3 = z_4 = z_5^3 - z_2^4 = z_6^3 - z_2^5 = 0$

and this does not necessarily imply that $z \in V$. Considering only the three-dimensional space of the coordinates (z_2, z_5, z_6), the parametric equations $z_2 = t^3$, $z_5 = t^4$, $z_6 = t^5$ describe a one-dimensional complex analytic subvariety $W \subseteq \mathbb{C}^3$ which appears as a three-sheeted branched analytic covering of the coordinate axis of the variable z_2 under the natural projection, while the equations $z_5^3 - z_2^4 = z_6^3 - z_2^5 = 0$ describe a one-dimensional complex analytic subvariety of $\mathbb{C}^3$ which appears as a nine-sheeted branched analytic covering of that axis under the same projection hence which contains W as a proper analytic subvariety; the last two equations are just the first set of canonical equations for the ideal of W, and to describe W precisely it is necessary to add the canonical equation of the second set $z_6 z_2 - z_5^2 = 0$. Therefore

$$V = \{z \in \mathbb{C}^6 \mid p_3(z) = p_4(z) = p_5(z) = p_6(z) = q_4(z)$$
$$= q_5(z) = q_6(z) = z_2 z_6 - z_5^2 = 0\} .$$

It should be pointed out though that upon interchanging the roles of the coordinates z_3 and z_5 the second set of canonical equations take the form

$$\tilde{q}_3(z_1, z_2; z_5, z_3) = z_2^8 z_3 - z_1 z_2^7 z_5 ,$$

$$\tilde{q}_4(z_1, z_2; z_5, z_4) = z_2^8 z_4 - z_1 z_2^6 z_5^2 ,$$

$$\tilde{q}_6(z_1, z_2; z_5, z_6) = z_2^8 z_6 - z_2^7 z_5^2 ;$$

and these together with the first set of canonical equations do serve to describe the subvariety V precisely; thus

$$V = \{z \in \mathbb{C}^6 | \; z_3^3 - z_1^3 z_2 = z_4^3 - z_1^3 z_2^2 = z_5^3 - z_2^4 = z_6^3 - z_2^5$$
$$= z_2 z_3 - z_1 z_5 = z_2^2 z_4 - z_1 z_5^2 = z_2 z_6 - z_5^2 = 0\} ,$$

but even though these reduced canonical equations now describe the subvariety V precisely they do not generate the full ideal $\text{id}\, V \subseteq {}_6\mathcal{O}$. To see this, note that if these functions generated the proper ideal of the subvariety V at the origin then by coherence they would also generate the proper ideal of the subvariety V at the regular points $\varphi(t_1,0) = (t_1,0,0,0,0,0) \in V$ for all sufficiently small nonzero values of the parameter t_1; but at any such point the Jacobian matrix consisting of the first partial derivatives of the seven reduced canonical equations has rank 2 rather than rank 4, hence these functions cannot generate the proper ideal of the variety V. In case (iii') with $b_{11} = b_{02} = 0$ the imbedding dimension of the corresponding variety is again six, and in the remaining examples the imbedding dimension exceeds six. These observations are summarized in Table 2; the blanks in that table are merely an avoidance of dull labor.

There are of course many germs of two-dimensional irreducible complex analytic varieties having regular normalizations but not having merely isolated singularities, and these correspond to equivalence classes of subalgebras $\mathcal{R} \subseteq {}_2\mathcal{O}$ which satisfy conditions (a), (b), (c) of Theorem 11 and its corollaries but which do not necessarily contain a power of the maximal ideal of

${}_2\mathcal{O}$. It is rather pointless to attempt here any systematic discussion of the general situation, but it may be of some interest to see a random illustrative example. For any positive integers m, n note that the subset $\mathcal{R} = \mathbb{C} + {}_1\mathcal{O}\cdot t_1^m + {}_2\mathcal{O}\cdot t_2^n \subseteq {}_2\mathcal{O}$ consisting of all germs of holomorphic functions $f \in {}_2\mathcal{O}$ of the form $f(t_1,t_2) = c + t_1^m g(t_1) + t_2^n h(t_1,t_2)$, for arbitrary $c \in \mathbb{C}$, $g \in {}_1\mathcal{O}$, $h \in {}_2\mathcal{O}$, is evidently a subalgebra of ${}_2\mathcal{O}$. Clearly the ideal $\mathfrak{A} = {}_2\mathcal{O}\cdot(\mathcal{R} \cap {}_2\mathfrak{m}) = {}_2\mathcal{O}\cdot t_1^m + {}_2\mathcal{O}\cdot t_2^n$ has $\operatorname{loc}\mathfrak{A} = 0$, so that the subalgebra $\mathcal{R}$ satisfies condition (a); an element $f \in {}_2\mathcal{O}$ having the Taylor expansion $f(t_1,t_2) = \Sigma_{\nu_1,\nu_2}\, c_{\nu_1\nu_2} t_1^{\nu_1} t_2^{\nu_2}$ belongs to $\mathcal{R}$ if and only if all coefficients c_{ν_1,ν_2} with $\nu_2 < n$ are zero except c_{00}, $c_{m+\nu,0}$ for $\nu = 0,1,2,\ldots$, hence it can be determined whether f belongs to $\mathcal{R}$ by examining the individual Taylor coefficients so that the subalgebra $\mathcal{R}$ satisfies condition (b); and ${}_2\mathcal{O}\cdot t_2^n \subseteq \mathcal{R}$ so that the subalgebra $\mathcal{R}$ also satisfies condition (c). Therefore there is a simple analytic mapping $\varphi\colon \mathbb{C}^2 \to V$ from $\mathbb{C}^2$ to a germ V of complex analytic variety such that ${}_V\mathcal{O} \cong \varphi^*({}_V\mathcal{O}) = \mathcal{R} \subseteq {}_2\mathcal{O}$. To determine this germ more explicitly note that $\dim {}_R\mathfrak{m}/{}_R\mathfrak{m}^2 = m(n+1)$, indeed that the elements $t_1^{m_1+\nu_1}$, $t_1^{\nu_1} t_2^{n+\nu_2}$ for $\nu_1 = 0,1,\ldots,m-1$, $\nu_2 = 0,1,\ldots,n-1$, represent a basis for the complex vector space ${}_R\mathfrak{m}/{}_R\mathfrak{m}^2$; hence by Corollary 1 to Theorem 11 the finite analytic mapping $\varphi\colon \mathbb{C}^2 \to \mathbb{C}^{m(n+1)}$ defined by $\varphi(t_1,t_2) = \{t_1^{m+\nu_1}, t_1^{\nu_1} t_2^{n+\nu_2}\}$ $\nu_1 = 0,1,\ldots,m-1$, $\nu_2 = 0,1,\ldots,n-1$, has as its image the desired analytic variety V. If $m = n = 1$ then $\mathcal{R} = {}_2\mathcal{O}$ and the variety

Table 2

Germs of two-dimensional irreducible complex analytic varieties V with normalization $\varphi: \mathbb{C}^2 \to V$ such that ${}_2\mathfrak{m}^3 \subseteq \varphi^*({}_V\mathcal{O}) \subseteq {}_2\mathcal{O}$. (Column 1: local ring $\mathcal{R} = \varphi^*({}_V\mathcal{O}) \subseteq {}_2\mathcal{O}$; column 2: imbedding dimension of V; column 3: reference to the preceding discussion; column 4: parametrization by the normalization φ; column 5: defining equations for V.)

1: $\mathcal{R}$ =	2	3	4: $\varphi(t_1,t_2)$ =	5: defining equations
${}_2\mathcal{O}$	2	(vi)	(t_1,t_2)	regular analytic variety
$\mathbb{C}+\mathbb{C}t_1+{}_2\mathfrak{m}^2$	4	(v)	(t_1,t_2^2,t_2^3,t_1t_2)	$z_3^2=z_2^3,\ z_4^2=z_1^2z_2,$ $z_1z_3=z_2z_4$
$\mathbb{C}+\mathbb{C}t_1+\mathbb{C}t_1^2$ $+\ \mathbb{C}t_2^2+{}_2\mathfrak{m}^3$	4	(iv')	$(t_1,t_2^2,t_2^3,t_1^2t_2)$	$z_3^2=z_2^3,\ z_4^2=z_1^4z_2,$ $z_1^2z_3=z_2z_4$
$\mathbb{C}+\mathbb{C}t_1+\mathbb{C}t_1^2$ $+\mathbb{C}t_1t_2+{}_2\mathfrak{m}^3$	6	(iv')	$(t_1,t_2^3,t_2^4,t_2^5,$ $t_1t_2,t_1t_2^2)$	$z_3^3=z_2^4,\ z_4^3=z_2^5,\ z_5^3=z_1^3z_2,$ $z_6^3=z_1^3z_2^2,\ z_2z_4=z_3^2,$ $z_2z_5=z_1z_3,\ z_2^2z_6=z_1z_3^2$
$\mathbb{C}+\mathbb{C}t_1+\mathbb{C}t_1^2$ $+\ {}_2\mathfrak{m}^3$	6	(iii')	$(t_1,t_2^3,t_2^4,t_2^5,$ $t_1^2t_2,t_1t_2^2)$	$z_3^3=z_2^4,\ z_4^3=z_2^5,\ z_5^3=z_1^6z_2,$ $z_6^3=z_1^3z_2^2,\ z_2z_4=z_3^2,$ $z_2z_5=z_1^2z_3,\ z_2^2z_6=z_1z_3^2$
$\mathbb{C}+{}_2\mathfrak{m}^2$	7	(iv")	$(t_1^2,t_2^3,t_1t_2,t_2^2,$ $t_1^3,t_1^2t_2,t_1t_2^2)$	$z_3^6=z_1^3z_2^2,\ z_4^3=z_2^2,\ z_5^2=z_1^3,$ $z_6^3=z_1^3z_2,\ z_7^6=z_1^3z_2^4,$ $z_1z_4=z_3^2,\ z_2z_5=z_3^3,$ $z_1z_2z_6=z_3^4,\ z_1^2z_2z_7=z_3^5$

Table 2

(Continued)

1: $\mathfrak{R}$ =	2	3	4: $\varphi(t_1,t_2)$ =	5: defining equations
$\mathbb{C} + \mathbb{C}t_1^2 + \mathbb{C}t_2^2 + {}_2\mathfrak{m}^3$	8	(iii")	$(t_1^2, t_2^3, t_1t_2^2, t_2^2, t_1^3, t_1^2t_2, t_1^3t_2, t_1t_2^3)$	$z_3^6 = z_1^3z_2^4,\ z_4^3 = z_2^2,\ z_5^2 = z_1^3,$ $z_6^3 = z_1^3z_2,\ z_7^6 = z_1^9z_2^2,$ $z_8^2 = z_1z_2^2,\ z_1^2z_2^2z_4 = z_3^4,$ $z_2^2z_5 = z_3^3,\ z_2z_6 = z_3^2,$ $z_1z_2^3z_7 = z_3^5,\ z_1z_2z_8 = z_3^3$
$\mathbb{C} + \mathbb{C}t_1^2 + \mathbb{C}t_1t_2 + {}_2\mathfrak{m}^3$	9	(iii")	$(t_1^2, t_2^3, t_1t_2, t_1^3, t_1^2t_2, t_1t_2^2, t_1t_2^3, t_2^4, t_2^5)$	$z_3^6 = z_1^3z_2^2,\ z_4^2 = z_1^3,$ $z_5^3 = z_1^3z_2,\ z_6^6 = z_1^3z_2^4,$ $z_7^2 = z_1z_2^2,\ z_8^3 = z_2^4,\ z_9^3 = z_2^5,$ $z_2z_4 = z_3^3,\ z_1z_2z_5 = z_3^4,$ $z_1^2z_2z_6 = z_3^5,\ z_1z_7 = z_3^3,$ $z_1^2z_8 = z_3^4,\ z_1z_9 = z_2z_3^2,$ $z_2z_9 = z_8^2$
$\mathbb{C} + \mathbb{C}t_1^2 + {}_2\mathfrak{m}^3$	11	(ii)	$(t_1^2, t_1^3, t_1^2t_2, t_1t_2^2, t_2^3, t_1^3t_2, t_1^2t_2^2, t_1t_2^3, t_2^4, t_1t_2^4, t_2^5)$	*
$\mathbb{C} + \mathbb{C}(t_1^2 + t_2^2) + {}_2\mathfrak{m}^3$	11	(ii)	$(t_1^2 + t_2^2, t_1^3, t_1^2t_2, t_1t_2^2, t_2^3, t_1^4, t_1^3t_2^2, t_1t_2^3, t_2^4, t_1^5, t_2^5)$	*
$\mathbb{C} + {}_2\mathfrak{m}^3$	15	(i)	*	*

V is nonsingular, but otherwise V necessarily has singularities. The next simplest case is that in which $m = 1$, $n = 2$, and V has imbedding dimension 3; the variety V is described parametrically by $\varphi(t_1,t_2) = (t_1,t_2^2,t_2^3)$ or by its canonical equation $z_3^2 = z_2^3$, and is merely the product of a nonsingular one-dimensional variety and a singular one-dimensional variety. There are two cases in which V has imbedding dimension 4, the cases $m = 1$, $n = 3$ and $m = 2$, $n = 1$. In the first of these the variety V is described parametrically by $\varphi(t_1,t_2) = (t_1,t_2^3,t_2^4,t_2^5)$ or by its canonical equations $z_3^3 = z_2^4$, $z_4^3 = z_2^5$, $z_2z_4 = z_3^8$, and is again the product of a nonsingular one-dimensional variety and a singular one-dimensional variety. In the second of these cases the variety V is described parametrically by $\varphi(t_1,t_2) = (t_1^2,t_2,t_1^3,t_1t_2)$ or by its canonical equations $z_3^2 = z_1^3$, $z_4^2 = z_1z_2^2$, $z_1z_4 = z_2z_3$; this variety has merely an isolated singularity at the origin, indeed is case (v) in Table 2. There is one case in which V has imbedding dimension 5, but then again V is the product of a nonsingular one-dimensional variety and a singular one-dimensional variety. There are three cases in which V has imbedding dimension 6, the cases $m = 1$, $n = 5$ and $m = 3$, $n = 1$ and $m = 2$, $n = 2$; the first of these is another product of a nonsingular one-dimensional variety and a singular one-dimensional variety, the second is the isolated singularity of case (iv') in Table 2, and the third is the first really interesting case. In this last case, for $m = n = 2$, the corresponding variety V is described parametrically by $\varphi(t_1,t_2) = (t_1^2,t_2^3,t_1t_2^2,t_2^2,t_1^3,t_1t_2^4)$ or by the canonical equations $z_3^6 = z_1^3 z_2^4$, $z_4^3 = z_2^2$, $z_5^2 = z_1^3$, $z_6^6 = z_1^3 z_2^8$, $z_1^2 z_2^2 z_4 = z_3^4$, $z_2^2 z_5 = z_3^3$, $z_1^2 z_2^2 z_6 = z_3^5$. Since the subalgebra $\mathcal{R} \subseteq {}_2\mathcal{O}$ does not contain the functions $t_1^\nu t_2$ for any $\nu \geq 0$ it follows that $\mathcal{R}$ cannot contain any power of the maximal ideal of ${}_2\mathcal{O}$, hence V cannot have just an isolated singularity; but since $t_2^2 \cdot {}_2\mathcal{O} \subseteq \mathcal{R}$ the function t_2^2 is a universal denominator for V, hence wherever. $t_2 \neq 0$ the variety V must be equivalent to its normalization and therefore nonsingular. The subvariety of V defined by the vanishing of the function t_2 is the one-dimensional irreducible subvariety

$$W = \{z \in V \mid z_2 = 0\} = \{z \in \mathbb{C}^6 \mid z_5^2 = z_1^3,\ z_2 = z_3 = z_4 = z_6 = 0\} ;$$

and since $0 \subset \mathcal{S}(V) \subseteq W$ it then follows that $\mathcal{S}(V) = W$.

§3. Finite analytic mappings with given range.

(a) Consider next the problem of describing all finite analytic mappings from germs of complex analytic varieties to a given germ V of a complex analytic variety. If $\varphi: W \to V$ is a finite analytic mapping the induced homomorphism $\varphi^*: {}_V\mathcal{O} \to {}_W\mathcal{O}$ can be viewed as exhibiting ${}_W\mathcal{O}$ as a finitely generated ${}_V\mathcal{O}$-module; conversely if ${}_W\mathcal{O}$ has the structure of a finitely generated ${}_V\mathcal{O}$-module then the mapping $\varphi^*: {}_V\mathcal{O} \to {}_W\mathcal{O}$ defined by $\varphi^*(f) = f\cdot 1 \in {}_W\mathcal{O}$ is clearly a finite homomorphism of complex algebras preserving the identities, and by Theorem 3(b) this is the homomorphism induced by a finite analytic mapping $\varphi: W \to V$. Thus the problem of interest here can be reduced to the more algebraic problem of describing the complex algebras with identities which are finitely generated ${}_V\mathcal{O}$-modules and which are isomorphic to the local rings of complex analytic varieties, although of course this will only be of interest if the property that the algebras be isomorphic to the local rings of complex analytic varieties can be replaced by some simpler and more purely algebraic properties. It is not sufficient merely to require that these algebras have no nilpotent elements; for example the formal algebraic extension ${}_1\mathcal{O}[r]$ be an element r such that $r^2 = 1$ is a finitely generated ${}_1\mathcal{O}$-module with no nilpotent elements, but it cannot be the local ring of a complex analytic variety since then $r^2 = 1$ would imply that $r = \pm 1 \in {}_1\mathcal{O}$. It is sufficient however to require that these algebras be local rings with no nilpotent elements.

<u>Theorem 14</u>. For any germ V of a complex analytic variety, the ${}_V\mathcal{O}$-modules which correspond to finite analytic mappings from germs of complex analytic varieties to V are precisely the finitely generated ${}_V\mathcal{O}$-modules which are also local rings with no nilpotent elements.

Proof. It is evidently only necessary to show that if a finitely generated ${}_V\mathcal{O}$-module is a local ring with no nilpotent elements then it is isomorphic to the local ring of some germ of complex analytic variety; indeed since there must exist a finite analytic mapping $V \rightarrow \mathbb{C}^k$ for some k and since under the induced homomorphism ${}_k\mathcal{O} \longrightarrow {}_V\mathcal{O}$ any finitely generated ${}_V\mathcal{O}$-module can also be viewed as a finitely generated ${}_k\mathcal{O}$-module, it is enough to consider only the special case of a regular germ V of a complex analytic variety. Thus suppose that $\mathcal{R}$ is a finitely generated ${}_k\mathcal{O}$-module which is also a local ring with maximal ideal ${}_{\mathcal{R}}\mathcal{M} \subseteq \mathcal{R}$; as usual the module structure can be viewed as that induced by a ring homomorphism $\varphi: {}_k\mathcal{O} \longrightarrow \mathcal{R}$, since $\mathcal{R}$ has an identity, and the homomorphism φ is the identity on the complex constants as they are naturally imbedded in ${}_k\mathcal{O}$ and in $\mathcal{R}$. As a preliminary observation note that $\varphi({}_k\mathcal{M}) \subseteq {}_{\mathcal{R}}\mathcal{M}$. Indeed if $f \in {}_k\mathcal{M}$ but $\varphi(f) \notin {}_{\mathcal{R}}\mathcal{M}$ there would be an element $r \in \mathcal{R}$ such that $r\cdot\varphi(f) = 1$. Since $\mathcal{R}$ is a finitely generated ${}_k\mathcal{O}$-module the element r would be integral over the subring $\varphi({}_k\mathcal{O}) \subseteq \mathcal{R}$, hence there would exist elements $f_i \in {}_k\mathcal{O}$ such that $0 = r^\nu + \varphi(f_1)r^{\nu-1} + \ldots + \varphi(f_\nu)$; and multiplying by $\varphi(f)^\nu$ it would follow that $0 = 1 + \varphi(f_1)\varphi(f) + \ldots + \varphi(f_\nu)\varphi(f^\nu) = \varphi(1 + f_1 f + \ldots + f_\nu f^\nu)$. The element $1 + f_1 f + \ldots + f_\nu f^\nu$ would then belong to the kernel of φ and hence to the maximal ideal ${}_k\mathfrak{m} \subset {}_k\mathcal{O}$, but that is impossible since $f \in {}_k\mathfrak{m}$. As another preliminary observation note that for any element $r \in \mathcal{R}$ there is a constant $c \in \mathbb{C}$ such that $r - c \in {}_{\mathcal{R}}\mathfrak{m}$. To see this, since r is integral over the subring $\varphi({}_k\mathcal{O}) \subseteq \mathcal{R}$ there are elements $f_i \in {}_k\mathcal{O}$ such that $0 = r^\nu + \varphi(f_1)r^{\nu-1} + \ldots + \varphi(f_\nu)$; and writing $f_i = a_i + \tilde{f}_i$ where $a_i \in \mathbb{C}$ is the constant term in the Taylor expansion of f_i and $\tilde{f}_i \in {}_k\mathfrak{m}$, and recalling from the preceding observation that $\varphi(\tilde{f}_i) \in \varphi({}_k\mathfrak{m}) \subseteq {}_{\mathcal{R}}\mathfrak{m}$, it follows that

$$r^\nu + a_1 r^{\nu-1} + \ldots + a_\nu = -\varphi(\tilde{f}_1)r^{\nu-1} - \ldots - \varphi(\tilde{f}_\nu) \in {}_{\mathcal{R}}\mathfrak{m} .$$

Letting $c_1,\ldots,c_\nu$ be the roots of the polynomial $X^\nu + a_1 X^{\nu-1} + \ldots + a_\nu$, then $(r - c_1)\cdots(r - c_\nu) = r^\nu + a_1 r^{\nu-1} + \ldots + a_\nu \in {}_{\mathcal{R}}\mathfrak{m}$; and since ${}_{\mathcal{R}}\mathfrak{m}$ is a prime ideal necessarily $r - c_i \in {}_{\mathcal{R}}\mathfrak{m}$ for some index i.

Now let $1,r_1,\ldots,r_n$ be any elements of $\mathcal{R}$ which generate $\mathcal{R}$ as an ${}_k\mathcal{O}$-module; as a consequence of the last observation above it can be assumed that $r_i \in {}_{\mathcal{R}}\mathfrak{m}$. The homomorphism $\varphi\colon {}_k\mathcal{O} \longrightarrow \mathcal{R}$ can be extended to a homomorphism $\Phi\colon {}_k\mathcal{O}[X_1,\ldots,X_n] \longrightarrow \mathcal{R}$ from the polynomial ring in indeterminates $X_1,\ldots,X_n$ over ${}_k\mathcal{O}$ to the ring $\mathcal{R}$ by defining $\Phi(P) = \Sigma_\nu\, \varphi(f_{\nu_1\ldots\nu_n})r_1^{\nu_1}\cdots r_n^{\nu_n}$ for any polynomial $P = \Sigma_\nu\, f_{\nu_1\ldots\nu_n}X_1^{\nu_1}\cdots X_n^{\nu_n} \in {}_k\mathcal{O}[X_1,\ldots,X_n]$; it is obvious that this is a surjective homomorphism, so if the ideal

$\mathfrak{A} \subseteq {}_k\mathcal{O}[X_1,\ldots,X_n]$ is the kernel of Φ then ${}_k\mathcal{O}[X_1,\ldots,X_n]/\mathfrak{A} \cong \mathcal{R}$. The polynomial ring ${}_k\mathcal{O}[X_1,\ldots,X_n]$ can be viewed as a subring of the ring ${}_{k+n}\mathcal{O}$ of germs of holomorphic functions of $k+n$ complex variables, and the ideal $\mathfrak{A} \subseteq {}_k\mathcal{O}[X_1,\ldots,X_n]$ generates an ideal $\tilde{\mathfrak{A}} = {}_{k+n}\mathcal{O}\cdot\mathfrak{A} \subseteq {}_{k+n}\mathcal{O}$. Note that if a polynomial $P \in \mathfrak{A}$ has constant term $f_{0\ldots0} \in {}_k\mathcal{O}$ then $0 = \Phi(P) = \varphi(f_{0\ldots0}) + r$ where $r \in \mathcal{R}r_1 + \ldots + \mathcal{R}r_n \subseteq {}_{\mathcal{R}}\mathfrak{m}$ since by assumption $r_i \in {}_{\mathcal{R}}\mathfrak{m}$; thus $\varphi(f_{0\ldots0}) \in \varphi({}_k\mathcal{O}) \cap {}_{\mathcal{R}}\mathfrak{m} \subseteq \varphi({}_k\mathfrak{m})$, and consequently $f_{0\ldots0} \in {}_k\mathfrak{m}$. When viewed as a holomorphic function of $k+n$ complex variables the polynomial P must therefore belong to the maximal ideal ${}_{k+n}\mathfrak{m}$; thus $\mathfrak{A} \subseteq {}_{k+n}\mathfrak{m}$, and $\tilde{\mathfrak{A}} \subseteq {}_{k+n}\mathfrak{m}$ is a proper ideal. The inclusions ${}_k\mathcal{O}[X_1,\ldots,X_n] \longrightarrow {}_{k+n}\mathcal{O}$ and $\mathfrak{A} \longrightarrow \tilde{\mathfrak{A}}$ induce a homomorphism $\Theta: {}_k\mathcal{O}[X_1,\ldots,X_n]/\mathfrak{A} \longrightarrow {}_{k+n}\mathcal{O}/\tilde{\mathfrak{A}}$, and to conclude the proof it suffices to show that Θ is an isomorphism; for in that case $\mathcal{R} \cong {}_k\mathcal{O}[X_1,\ldots,X_n]/\mathfrak{A} \cong {}_{k+n}\mathcal{O}/\tilde{\mathfrak{A}}$, and if $\mathcal{R}$ contains no nilpotent elements then $\tilde{\mathfrak{A}}$ is a radical ideal, $\tilde{\mathfrak{A}} = \text{id}\, W$ where $W = \text{loc}\,\tilde{\mathfrak{A}}$, and hence $\mathcal{R} \cong {}_W\mathcal{O}$. First since $r_i \in \mathcal{R}$ is integral over the subring $\varphi({}_k\mathcal{O}) \subseteq \mathcal{R}$ there is a monic polynomial $P_i \in \mathfrak{A} \cap {}_k\mathcal{O}[X_i]$ exhibiting this integral dependence; as an element of ${}_{k+1}\mathcal{O}$ the function $P_i \in \mathfrak{A} \subseteq {}_{k+n}\mathfrak{m}$ is regular in X_i, so by the Weierstrass preparation theorem $P_i = U_i\tilde{P}_i$ where U_i is a unit in ${}_{k+n}\mathcal{O}$ and $\tilde{P}_i \in {}_k\mathcal{O}[X_i]$ is a Weierstrass polynomial. Then for any element $F \in {}_{k+n}\mathcal{O}$ repeated use of the Weierstrass division theorem in a familiar manner shows that $F = \Sigma_i \tilde{P}_i G_i + P$ where $G_i \in {}_{k+n}\mathcal{O}$ and $P \in {}_k\mathcal{O}[X_1,\ldots,X_n]$; thus

$F \in \tilde{\mathfrak{A}} + {}_k\mathcal{O}[X_1,\ldots,X_n]$, since $\tilde{P}_i = U_i^{-1}P_i \in \tilde{\mathfrak{A}}$, and it is clear from this that the homomorphism Θ is surjective. Second consider a polynomial $P \in {}_k\mathcal{O}[X_1,\ldots,X_n]$ which represents an element in the kernel of the homomorphism Θ, that is, a polynomial $P \in \tilde{\mathfrak{A}} \cap {}_k\mathcal{O}[X_1,\ldots,X_n]$. If $P_1,\ldots,P_m$ are generators of the ideal $\mathfrak{A}$ in ${}_k\mathcal{O}[X_1,\ldots,X_n]$, hence also generators of the ideal $\tilde{\mathfrak{A}}$ in ${}_{k+n}\mathcal{O}$, then $P = \Sigma_i F_iP_i$ for some elements $F_i \in {}_{k+n}\mathcal{O}$. Now for any positive integer N the Taylor expansion of the function F_i can be split into a sum $F_i = F_i^N + \tilde{F}_i^N$ where $F_i^N \in {}_k\mathcal{O}[X_1,\ldots,X_n]$ is a polynomial of degree at most N in the variables $X_1,\ldots,X_n$ and each monomial in the Taylor series $\tilde{F}_i^N$ is of degree greater than N in the variables $X_1,\ldots,X_n$; then $P - \Sigma_i F_i^NP_i = \Sigma_i \tilde{F}_i^NP_i$ where the left hand side is a polynomial in ${}_k\mathcal{O}[X_1,\ldots,X_n]$ and each monomial in the Taylor series expansion of the right hand side is a polynomial of degree greater than N in the variables $X_1,\ldots,X_n$, hence where each monomial in the polynomial $P - \Sigma_i F_i^NP_i$ must be of degree greater than N in the variables $X_1,\ldots,X_n$. Applying the homomorphism Φ and noting that $F_i^NP_i \in \mathfrak{A}$ and $\Phi(X_i) = r_i \in {}_{\mathcal{R}}\mathfrak{m}$, it follows that $\Phi(P) = \Phi(P - \Sigma_i F_i^NP_i) \in {}_{\mathcal{R}}\mathfrak{m}^N$; but since this holds for any positive integer N and $\mathcal{R}$ is by assumption a local ring, it further follows from Nakayama's lemma that $\Phi(P) = 0$, hence that $P \in \mathfrak{A}$. The homomorphism Θ is therefore injective, hence is an isomorphism; and as noted, that suffices to conclude the proof.

A restatement of the essential content of the preceding theorem in the following form is also useful.

Corollary 1 to Theorem 14. The local rings of germs of complex analytic varieties of dimension at most k are precisely the finitely generated ${}_k\mathcal{O}$-modules which are local rings with no nilpotent elements.

It is apparent from the proof of the theorem that the restriction of having no nilpotents can be dropped, in the sense that the finitely generated ${}_k\mathcal{O}$-modules which are local rings are precisely the rings of the form ${}_n\mathcal{O}/\mathfrak{A}$ for some ideal $\mathfrak{A} \subseteq {}_n\mathcal{O}$ but not necessarily a radical ideal; these rings are the local rings of generalized or nonreduced complex analytic varieties, which arise in many contexts but which have not been and will not be considered at present. As usual the situation is somewhat simpler when only irreducible complex analytic varieties are considered; in both Theorem 14 and its Corollary 1 the hypothesis that the ring be a local ring can be dropped in case that ring has no zero divisors.

Corollary 2 to Theorem 14. For any germ V of a complex analytic variety, the ${}_V\mathcal{O}$-modules which correspond to finite analytic mappings from irreducible germs of complex analytic varieties to V are precisely the finitely generated ${}_V\mathcal{O}$-modules which are also integral domains with identities. In particular, the local rings of irreducible germs of complex analytic varieties of dimension at most k can be characterized as precisely the finitely generated ${}_k\mathcal{O}$-modules which are integral domains with identities.

Proof. In view of Theorem 14 and its Corollary 1 it is clearly only necessary to show that a finitely generated ${}_k\mathcal{O}$-module $\mathcal{R}$ which is an integral domain with an identity is also a local ring; $\mathcal{R}$ is of course Noetherian since ${}_k\mathcal{O}$ is, so what remains is to show that the nonunits of $\mathcal{R}$ form an ideal. Since $\mathcal{R}$ contains an identity the module structure on $\mathcal{R}$ can be viewed as that induced by a ring homomorphism $\varphi\colon {}_k\mathcal{O} \longrightarrow \mathcal{R}$. Note that the ideal $\varphi({}_k\mathfrak{m})\cdot\mathcal{R} \subseteq \mathcal{R}$ is necessarily a proper ideal as a consequence of Nakayama's Lemma, since $\mathcal{R}$ is a finitely generated module over the local ring ${}_k\mathcal{O}$; this ideal must be contained in some maximal ideal ${}_{\mathcal{R}}\mathfrak{m}$ of $\mathcal{R}$, and the proof will be completed by showing that all nonunits of $\mathcal{R}$ are contained in ${}_{\mathcal{R}}\mathfrak{m}$.

Any element $r \in \mathcal{R}$ is integral over the subring $\varphi({}_k\mathcal{O}) \subseteq \mathcal{R}$, hence there are germs $f_i \in {}_k\mathcal{O}$ such that $r^\nu + \varphi(f_1)r^{\nu-1} + \ldots + \varphi(f_\nu) = 0$. If $f_\nu \notin {}_k\mathfrak{m}$ then $1 = r\cdot\varphi(f_\nu^{-1})\cdot(-r^{\nu-1} - \varphi(f_1)r^{\nu-2} - \ldots - \varphi(f_{\nu-1}))$ so that r is a unit in $\mathcal{R}$; consequently if r is a nonunit in $\mathcal{R}$ then $f_\nu \in {}_k\mathfrak{m}$ and the polynomial $P(X) = X^\nu + f_1X^{\nu-1} + \ldots + f_\nu \in {}_k\mathcal{O}[X]$ is a nonunit in ${}_{k+1}\mathcal{O}$ when viewed as the germ of a holomorphic function of $k+1$ complex variables. It then follows from the Weierstrass preparation and division theorems that $P(X) = U(X)\cdot\tilde{P}(X)$ where $U(X) \in {}_k\mathcal{O}[X]$ is a unit in ${}_{k+1}\mathcal{O}$ and $\tilde{P}(X) \in {}_k\mathcal{O}[X]$ is a Weierstrass polynomial. Letting $\varphi P(X) = X^\nu + \varphi(f_1)X^{\nu-1} + \ldots + \varphi(f_\nu) \in \varphi({}_k\mathcal{O})[X]$, note that $0 = \varphi P(r) = \varphi U(r)\cdot\varphi\tilde{P}(r)$; but since r is a nonunit then as above it is impossible that $\varphi U(r) = 0$, hence since $\mathcal{R}$ is by hypothesis an integral domain necessarily $\varphi\tilde{P}(r) = 0$. Therefore if r is a nonunit it can be assumed that $P(X) \in {}_k\mathcal{O}[X]$ is a Weierstrass polynomial, hence that $f_i \in {}_k\mathfrak{m}$ for $i = 1,\ldots,\nu$; and then

$$r^{\nu} = -\varphi(f_1)r^{\nu-1} - \ldots - \varphi(f_{\nu}) \in \varphi({}_k\mathfrak{m})\cdot\mathcal{R} \subseteq {}_{\mathcal{R}}\mathfrak{m},$$

so since ${}_{\mathcal{R}}\mathfrak{m}$ is a prime ideal necessarily $r \in {}_{\mathcal{R}}\mathfrak{m}$ and the proof is thereby concluded.

Some further applications of these results which will not be pursued here can be found in the paper by A. Seidenberg, Saturation of an analytic ring, American Journal of Math. vol. 94 (1972), pp. 424-430; the proofs of Theorem 14 and its corollaries were adapted from this paper.

(b) For any germ V of complex analytic analytic variety of dimension at most k there is a finite analytic mapping $\varphi: V \to \mathbb{C}^k$ which exhibits the local ring ${}_V\mathcal{O}$ as a finitely generated ${}_k\mathcal{O}$-module. This is of relatively little immediate use in attempting to classify germs of complex analytic varieties however, since it is quite obvious that analytic equivalences of germs of complex analytic varieties need not determine homomorphisms of ${}_k\mathcal{O}$-modules; indeed it is not apparent without some further thought just to what extent the properties of a local ring ${}_V\mathcal{O}$ as an ${}_k\mathcal{O}$-module are independent of the particular finite analytic mapping $\varphi: V \to \mathbb{C}^k$ inducing that module structure. The aim of the subsequent discussion is to examine this question and also to look into the analytic significance of this module structure. As motivation it is perhaps of interest to consider first the analytic significance of this module structure in the simplest case.

A germ V of complex analytic variety is said to be <u>perfect</u> if there is a finite analytic mapping $\varphi: V \to \mathbb{C}^k$ which exhibits the local ring ${}_V\mathcal{O}$ as a free ${}_k\mathcal{O}$-module. A regular germ of complex analytic variety is of course trivially perfect. It follows from Corollary 1 to Theorem 19 of CAV I that a pure-dimensional germ of a complex analytic variety which can be represented by the germ of a complex analytic subvariety of codimension one in $\mathbb{C}^{k+1}$ is also perfect; indeed if $\varphi: V \to U$ is a branched analytic covering of order r induced by the natural projection mapping $\mathbb{C}^{k+1} \longrightarrow \mathbb{C}^k$ when the germ V is represented by a complex analytic subvariety of an open neighborhood of the origin in $\mathbb{C}^{k+1}$ then the induced homomorphism $\varphi^*: {}_k\mathcal{O} \longrightarrow {}_V\mathcal{O}$ exhibits ${}_V\mathcal{O}$ as an ${}_k\mathcal{O}$-module isomorphic to ${}_k\mathcal{O}^r$. As yet another example, any irreducible one-dimensional germ of a complex analytic variety is necessarily perfect. To see this, for any irreducible one-dimensional germ V choose any finite analytic mapping $\varphi: V \to \mathbb{C}^1$ which represents V as an r-sheeted branched analytic covering, and let $\mathfrak{A} \subseteq {}_V\mathcal{O}$ be the characteristic ideal of the mapping φ; it follows from Corollary 1 to Theorem 7 that $\dim_{\mathbb{C}}({}_V\mathcal{O}/\mathfrak{A}) \geq r$ with equality holding precisely when ${}_V\mathcal{O}$ is a free ${}_1\mathcal{O}$-module, so to show that V is perfect it is enough to show that $\dim_{\mathbb{C}}({}_V\mathcal{O}/\mathfrak{A}) = r$. Let $\rho: \mathbb{C}^1 \to V$ be the normalization of V, so

that $\varphi \circ \rho\colon \mathbb{C}^1 \to \mathbb{C}^1$ is an r-sheeted branched analytic covering which in terms of suitable local coordinates is just the mapping $(\varphi \circ \rho)(t) = t^r$; the local ring ${}_V\mathcal{O}$ is isomorphic to its image $\rho^*({}_V\mathcal{O}) = \mathcal{R} \subseteq {}_1\mathcal{O}$, and under this isomorphism the characteristic ideal $\mathfrak{A} \subseteq {}_V\mathcal{O}$ is evidently transformed into the subset $\rho^*(\mathfrak{A}) = \mathcal{R}\cdot t^r \subseteq \mathcal{R} \subseteq {}_1\mathcal{O}$. Now considering the vector spaces $\mathcal{R}\cdot t^r \subseteq {}_1\mathcal{O}\cdot t^r \subseteq {}_1\mathcal{O}$ it follows that
$\dim_{\mathbb{C}}({}_1\mathcal{O}/\mathcal{R}\cdot t^r) = \dim_{\mathbb{C}}({}_1\mathcal{O}\cdot t^r/\mathcal{R}\cdot t^r) + \dim_{\mathbb{C}}({}_1\mathcal{O}/{}_1\mathcal{O}\cdot t^r) =$
$\dim_{\mathbb{C}}({}_1\mathcal{O}/\mathcal{R}) + r < \infty$, and considering the vector spaces $\mathcal{R}\cdot t^r \subseteq \mathcal{R} \subseteq {}_1\mathcal{O}$ it follows that
$\dim_{\mathbb{C}}({}_1\mathcal{O}/\mathcal{R}\cdot t^r) = \dim_{\mathbb{C}}(\mathcal{R}/\mathcal{R}\cdot t^r) + \dim_{\mathbb{C}}({}_1\mathcal{O}/\mathcal{R}) < \infty$; and upon comparing these two equalities it then follows that
$\dim_{\mathbb{C}}({}_V\mathcal{O}/\mathfrak{A}) = \dim_{\mathbb{C}}(\mathcal{R}/\mathcal{R}\cdot t^r) = r$ as desired. Note incidentally that ${}_V\mathcal{O}$ is thus exhibited as a free ${}_1\mathcal{O}$-module by any finite analytic mapping $\varphi\colon V \to \mathbb{C}^1$.

On the other hand not all germs of complex analytic varieties are perfect; for example, it is easy to see that a perfect germ of complex analytic variety is necessarily pure-dimensional. Indeed if $\varphi\colon V \to \mathbb{C}^k$ is a finite analytic mapping exhibiting ${}_V\mathcal{O}$ as a free ${}_k\mathcal{O}$-module and $f \in {}_k\mathcal{O}$ is nonzero then $\varphi^*(f)$ cannot be a zero divisor in ${}_V\mathcal{O}$; but if there were an irreducible component $V_1 \subseteq V$ with $\dim V_1 < k$ then the image $\varphi(V_1)$ would be the germ of a proper complex analytic subvariety in $\mathbb{C}^k$, and for any nonzero $f \in {}_k\mathcal{O}$ vanishing on $\varphi(V_1)$ clearly $\varphi^*(f)$ would be a zero divisor in ${}_V\mathcal{O}$. Actually perfect germs are a quite restrictive class of germs of complex analytic varieties and have a number of interesting special properties. To introduce a convenient terminology for describing one of these properties, a subset W of a complex analytic variety V is called a <u>removable singularity for holomorphic functions</u> if every holomorphic function on $V - W$ extends to a holomorphic function on V, or equivalently, if the natural restriction mapping $\Gamma(V, {}_V\mathcal{O}) \rightarrow \Gamma(V - W, {}_V\mathcal{O})$ is an isomorphism. The extended Riemann removable singularities theorem can then be rephrased as the assertion that on a connected k-dimensional complex analytic manifold any complex analytic subvariety W for which $\dim W \leq k - 2$ is a removable singularity for holomorphic functions; and Corollary 1 to Theorem 12 can be rephrased as the assertion that on a connected k-dimensional normal complex analytic variety any complex analytic subvariety W for which $\dim W \leq k - 2$ is a removable singularity for holomorphic functions. The analogous assertion cannot be made even for a pure-dimensional but otherwise arbitrary complex analytic variety, since for example it is evidently false on a nonnormal variety V for which $\dim \mathcal{S}(V) \leq \dim V - 2$; but it can be made for perfect germs.

<u>Theorem 15</u>. On a perfect germ V of a complex analytic variety any complex analytic subvariety $W \subset V$ with $\dim W \leq \dim V - 2$ is a removable singularity for holomorphic functions.

Proof. Represent the germ V by a branched analytic covering $\varphi: V \rightarrow \mathbb{C}^k$ exhibiting ${}_V\mathcal{O}$ as a free ${}_k\mathcal{O}$-module of rank r; and note that if $W \subset V$ is a complex analytic subvariety with $\dim W \leq \dim V - 2 = k - 2$ then its image $\varphi(W)$ is a complex analytic subvariety of an open neighborhood of the origin in $\mathbb{C}^k$ with $\dim \varphi(W) \leq k - 2$ as well. Now the direct image sheaf $\varphi_*({}_V\mathcal{O})$ is a coherent analytic sheaf in an open neighborhood of the origin in $\mathbb{C}^k$, and at the origin itself the stalk $\varphi_*({}_V\mathcal{O})_0$ is isomorphic to ${}_k\mathcal{O}^r_0$; it then follows easily from coherence that the sheaves $\varphi_*({}_V\mathcal{O})$ and ${}_k\mathcal{O}^r$ are isomorphic analytic sheaves in a sufficiently small open neighborhood U of the origin in $\mathbb{C}^k$. This isomorphism in turn induces an isomorphism of the modules of sections of these sheaves, so that $\Gamma(V,{}_V\mathcal{O}) \cong \Gamma(U,\varphi_*({}_V\mathcal{O})) \cong \Gamma(U,{}_k\mathcal{O}^r)$; and this isomorphism commutes with the natural restriction mappings to yield an isomorphism $\Gamma(V - \varphi^{-1}(\varphi(W)),{}_V\mathcal{O}) \cong \Gamma(U - \varphi(W),{}_k\mathcal{O}^r)$. However it follows from the extended Riemann removable singularities theorem that the restriction mapping is actually an isomorphism $\Gamma(U,{}_k\mathcal{O}^r) \cong \Gamma(U - \varphi(W),{}_k\mathcal{O}^r)$; and consequently on V the restriction mapping is also an isomorphism $\Gamma(V,{}_V\mathcal{O}) \cong \Gamma(V - \varphi^{-1}(\varphi(W)),{}_V\mathcal{O})$, and that concludes the proof of the theorem.

Corollary 1 to Theorem 15. A perfect germ V of a complex analytic variety is normal if and only if $\dim \mathcal{S}(V) \leq \dim V - 2$.

Proof. For an arbitrary normal germ V of complex analytic variety it was proved in Theorem 12 that $\dim \mathcal{S}(V) \leq \dim V - 2$. Conversely if V is a perfect germ of a complex analytic variety and $\dim \mathcal{S}(V) \leq \dim V - 2$ then it follows from Theorem 15 that $\mathcal{S}(V)$ is a removable singularity; consequently any weakly holomorphic function on V extends to a holomorphic function, so that

V is normal and the proof of the corollary is thus concluded.

It follows from this corollary that the singular two-dimensional germs of complex analytic varieties with isolated singularities as listed in Table 2 in §2(d) cannot be perfect germs, since they are not normal; and any number of further examples of germs which are not perfect can be constructed similarly. Since pure two-dimensional subvarieties in $\mathbb{C}^3$ are as noted above necessarily perfect, it is clear why the examples in Table 2 all have imbedding dimension at least four, except of course for the regular variety. With the earlier observation that hypersurfaces are necessarily perfect still in mind, it is perhaps worth restating Corollary 1 in the following more concrete special case.

Corollary 2 to Theorem 15. A hypersurface V of an open subset $U \subseteq \mathbb{C}^k$ (that is, a complex analytic subvariety $V \subset U$ of pure dimension $k-1$) is normal if and only if $\dim \mathcal{S}(V) \leq \dim V - 2$.

Turning from germs of varieties to varieties themselves, it is natural to say that a complex analytic variety V is perfect at a point $p \in V$ if the germ of V at the point p is perfect, and to say that a complex analytic variety V is perfect if it is perfect at each of its points. It is easy to see that if a variety is perfect at a point p it is perfect at all points in a full open neighborhood of p. Indeed if V is perfect at a point p there is, after restricting to a sufficiently small open neighborhood of that point, a representation $\varphi: V \to \mathbb{C}^k$ of that neighborhood as a branched analytic covering which exhibits ${}_V\mathcal{O}_p$ as a free module of rank r over the ring ${}_k\mathcal{O}_{\varphi(p)}$; and again the direct image sheaf $\varphi_*({}_V\mathcal{O})$ is then isomorphic to the free analytic sheaf ${}_k\mathcal{O}^r$ in a full open neighborhood of the point $\varphi(p)$ in $\mathbb{C}^k$. Now for every point $q \in V$ the local ring ${}_V\mathcal{O}_q$ as a module over the ring ${}_k\mathcal{O}_{\varphi(q)}$ is a direct summand of the module $\varphi_*({}_V\mathcal{O})_{\varphi(q)} \cong {}_k\mathcal{O}^r_{\varphi(q)}$; hence it is sufficient just to show that if there are submodules $\mathcal{R}$, $\mathcal{S}$ of the free ${}_k\mathcal{O}$-module ${}_k\mathcal{O}^r$ such that ${}_k\mathcal{O}^r = \mathcal{R} \oplus \mathcal{S}$ then $\mathcal{R}$ and $\mathcal{S}$ are themselves free ${}_k\mathcal{O}$-modules. Suppose that the vector-valued functions $F_1,\ldots,F_m \in {}_k\mathcal{O}^r$ generate the ${}_k\mathcal{O}$-submodule $\mathcal{R} \subseteq {}_k\mathcal{O}^r$ and the functions $G_1,\ldots,G_n \in {}_k\mathcal{O}^r$ generate $\mathcal{S}$. It can be assumed that the vectors $F_1(0),\ldots,F_m(0)$ are linearly independent, and also of course that the vectors $G_1(0),\ldots,G_n(0)$ are linearly independent; for if say $F_m(0)$ is linearly dependent on $F_1(0),\ldots,F_{m-1}(0)$, then the elements $F_1,\ldots,F_{m-1}$ generate a submodule $\mathcal{R}' \subseteq \mathcal{R}$ for which $\mathcal{R} = \mathcal{R}' + \mathcal{R} \cap {}_k\mathfrak{m}\cdot{}_k\mathcal{O}^r = \mathcal{R}' + {}_k\mathfrak{m}\cdot\mathcal{R}$, and it follows from Nakayama's lemma that $\mathcal{R}' = \mathcal{R}$. Since ${}_k\mathcal{O}^r/{}_k\mathfrak{m}\cdot{}_k\mathcal{O}^r \cong (\mathcal{R}/{}_k\mathfrak{m}\cdot\mathcal{R}) \oplus (\mathcal{S}/{}_k\mathfrak{m}\cdot\mathcal{S})$ as modules over the residue class ring ${}_k\mathcal{O}/{}_k\mathfrak{m} \cong C$, that is as complex vector spaces, it follows that the vectors $F_1(0),\ldots,F_m(0)$ and $G_1(0),\ldots,G_n(0)$ span vector subspaces $\mathbb{C}^m$ and $\mathbb{C}^n$ in $\mathbb{C}^r$ such that $\mathbb{C}^r = \mathbb{C}^m \oplus \mathbb{C}^n$ The square matrix $(F_1,\ldots,F_m,G_1,\ldots,G_n)$, where F_i, G_i are viewed as column vectors, is then a nonsingular matrix of holomorphic functions which establishes an isomorphism ${}_k\mathcal{O}^m \oplus {}_k\mathcal{O}^n \cong \mathcal{R} \oplus \mathcal{S}$ as desired. (Note by the way that this argument applies equally well to modules over the local ring ${}_V\mathcal{O}_p$ of any complex analytic variety V.)

(c) For an arbitrary germ V of a complex analytic variety a finite analytic mapping $\varphi: V \to \mathbb{C}^k$ exhibits ${}_V\mathcal{O}$ as a somewhat more complicated ${}_k\mathcal{O}$-module than just a free module, and to examine this situation a further analysis of the general structure of ${}_k\mathcal{O}$-modules is necessary. It is useful and not at all inconvenient first to consider more generally an arbitrary finitely generated module A over the local ring ${}_V\mathcal{O}$ of any germ V of a complex analytic variety. To any choice of generators $a_1,\ldots,a_r$ for the module A there is associated the surjective homomorphism of ${}_V\mathcal{O}$-modules $\sigma: {}_V\mathcal{O}^r \to A$ defined by $\sigma(f_1,\ldots,f_r) = f_1a_1 + \ldots + f_ra_r$; and conversely any surjective homomorphism of ${}_V\mathcal{O}$-modules $\sigma: {}_V\mathcal{O}^r \to A$ can be viewed as so arising from some choice of generators for the module A. If r is the minimal number of generators for the module A then the kernel of any surjective homomorphism $\sigma: {}_V\mathcal{O}^r \to A$ will be called a <u>syzygy module</u> for A.

<u>Theorem 16</u>. If A is a finitely generated module over the local ring ${}_V\mathcal{O}$ of a germ V of a complex analytic variety then all syzygy modules for A are isomorphic modules. The minimal number of generators for the module A is $r = \dim_{\mathbb{C}} A/{}_V\mathfrak{m}\cdot A$; and if $n \geq r$ and $\sigma: {}_V\mathcal{O}^n \to A$ is any surjective module homomorphism then there is an isomorphism $\theta: {}_V\mathcal{O}^n \to {}_V\mathcal{O}^r \oplus {}_V\mathcal{O}^{n-r}$ such that $\sigma\theta^{-1} = \sigma|{}_V\mathcal{O}^r$.

Proof. The residue class module $\tilde{A} = A/{}_V\mathfrak{m}\cdot A$ can be viewed as a finitely generated module over the residue class ring ${}_V\mathcal{O}/{}_V\mathfrak{m} \cong \mathbb{C}$, hence as a finite-dimensional complex vector space. If $a_1,\ldots,a_n$ are any generators of the ${}_V\mathcal{O}$-module A then the residue classes $\tilde{a}_1,\ldots,\tilde{a}_n$ are generators of the complex vector space $\tilde{A}$; and if $\dim_{\mathbb{C}} \tilde{A} = r$ then r of these residue classes, say $\tilde{a}_1,\ldots,\tilde{a}_r$, form a basis for the complex vector space $\tilde{A}$. The elements $a_1,\ldots,a_r$ generate a submodule $A_1 \subseteq A$, and since evidently $A_1 + {}_V\mathfrak{m}\cdot A = A$ it follows from Nakayama's lemma that $A_1 = A$; thus the minimal number of generators of the ${}_V\mathcal{O}$-module A is $r = \dim_{\mathbb{C}} A/{}_V\mathfrak{m}\cdot A$. Furthermore there are germs $h_{ij} \in {}_V\mathcal{O}$ such that $a_i = \Sigma_{j=1}^{r} h_{ij}a_j$ for $r+1 \leq i \leq n$; and the module homomorphism $\theta\colon {}_V\mathcal{O}^n \longrightarrow {}_V\mathcal{O}^n$ defined by $\theta(f_1,\ldots,f_n) = (g_1,\ldots,g_n)$, where

$$g_j = \begin{cases} f_j - \Sigma_{i=r+1}^{n} f_i h_{ij} & \text{for } \quad 1 \leq j \leq r, \\ f_j & \text{for } \quad r+1 \leq j \leq n, \end{cases}$$

is clearly an isomorphism for which $\sigma(f_1,\ldots,f_n) = \Sigma_{j=1}^{n} f_j a_j = \Sigma_{j=1}^{r} f_j a_j + \Sigma_{i=r+1}^{n} \Sigma_{j=1}^{r} f_i h_{ij} a_j = \sigma(g_1,\ldots,g_r,0,\ldots,0)$. Thus $\sigma\theta^{-1}(g_1,\ldots,g_n) = \sigma(f_1,\ldots,f_n) = \sigma(g_1,\ldots,g_r,0,\ldots,0)$. Finally if $a_1,\ldots,a_r$ and $b_1,\ldots,b_r$ are two minimal sets of generators of the ${}_V\mathcal{O}$-module A then $a_i = \Sigma_{j=1}^{r} h_{ij}b_j$ for some germs $h_{ij} \in {}_V\mathcal{O}$ and consequently $\tilde{a}_i = \Sigma_{j=1}^{r} h_{ij}(0)\tilde{b}_j$; and since $\tilde{a}_1,\ldots,\tilde{a}_r$ and $\tilde{b}_1,\ldots,\tilde{b}_r$ are two bases for the complex vector space $\tilde{A}$ it follows that the constant matrix $\{h_{ij}(0)\}$ is nonsingular hence that the matrix of holomorphic functions $\{h_{ij}\}$ determines an isomorphism of ${}_V\mathcal{O}$-modules $\theta: {}_V\mathcal{O}^r \longrightarrow {}_V\mathcal{O}^r$. This isomorphism clearly transforms the kernel of the homomorphism $\sigma_a: {}_V\mathcal{O}^r \longrightarrow A$ associated to the generators $a_1,\ldots,a_r$ into the kernel of the homomorphism $\sigma_b: {}_V\mathcal{O}^r \longrightarrow A$ associated to the generators $b_1,\ldots,b_r$; consequently all syzygy modules for A are isomorphic, and the proof of the theorem is thereby concluded.

This theorem can be restated as the assertions that to each finitely generated ${}_V\mathcal{O}$-module A there is associated a unique syzygy module, which will be denoted by syz A, such that there exists an exact sequence of ${}_V\mathcal{O}$-modules of the form

$$0 \longrightarrow \text{syz } A \xrightarrow{i} {}_V\mathcal{O}^r \xrightarrow{\sigma} A \longrightarrow 0$$

where $r = \dim_{\mathbb{C}} A/{}_V\mathfrak{m}\cdot A$; and that for any other exact sequence of ${}_V\mathcal{O}$-modules of the form

$$0 \longrightarrow A_1 \xrightarrow{j} {}_V\mathcal{O}^n \xrightarrow{\tau} A \longrightarrow 0$$

necessarily $n \geq r$ and there are isomorphisms $\theta: {}_V\mathcal{O}^n \longrightarrow {}_V\mathcal{O}^r \oplus {}_V\mathcal{O}^{n-r}$ and $\theta_1: A_1 \longrightarrow \text{syz } A \oplus {}_V\mathcal{O}^{n-r}$ such that the following diagram is commutative

$$\begin{array}{ccccccccc}
0 & \longrightarrow & A_1 & \xrightarrow{j} & {}_V\mathcal{O}^n & \xrightarrow{\tau} & A & \longrightarrow & 0 \\
 & & \downarrow{\scriptstyle \theta_1} & & \downarrow{\scriptstyle \theta} & & \downarrow{\scriptstyle 1} & & \\
0 & \longrightarrow & \text{syz } A \oplus {}_V\mathcal{O}^{n-r} & \xrightarrow{I} & {}_V\mathcal{O}^r \oplus {}_V\mathcal{O}^{n-r} & \xrightarrow{\Sigma} & A & \longrightarrow & 0
\end{array}$$

where $I(F,G) = (i(F),G)$ for any elements $F \in \text{syz}\, A$ and $G \in {}_V\mathcal{O}^{n-r}$ and $\Sigma\,(F,G) = \sigma(F)$ for any elements $F \in {}_V\mathcal{O}^{r}$ and $G \in {}_V\mathcal{O}^{n-r}$.

Corollary 1 to Theorem 16. If A and B are finitely generated modules over the local ring ${}_V\mathcal{O}$ of a germ V of a complex analytic variety then

$$\text{syz}\,(A \oplus B) \cong (\text{syz}\, A) \oplus (\text{syz}\, B)\ .$$

Proof. The direct sum of the exact sequences of ${}_V\mathcal{O}$-modules

$$0 \longrightarrow \text{syz}\, A \longrightarrow {}_V\mathcal{O}^{r} \longrightarrow A \longrightarrow 0$$

and

$$0 \longrightarrow \text{syz}\, B \longrightarrow {}_V\mathcal{O}^{s} \longrightarrow B \longrightarrow 0$$

is the exact sequence of ${}_V\mathcal{O}$-modules

$$0 \longrightarrow (\text{syz}\, A) \oplus (\text{syz}\, B) \longrightarrow {}_V\mathcal{O}^{r+s} \longrightarrow A \oplus B \longrightarrow 0;$$

and since it is clear that

$$\dim_{\mathbb{C}}\,(A \oplus B)/{}_V\mathfrak{m}\,(A \oplus B) = \dim_{\mathbb{C}}\, A/{}_V\mathfrak{m}A + \dim_{\mathbb{C}}\, B/{}_V\mathfrak{m}B = r + s$$

it follows immediately from Theorem 16 that $\text{syz}\,(A \oplus B) \cong (\text{syz}\, A) \oplus (\text{syz}\, B)$ as desired.

A module A is free precisely when $\text{syz}\, A = 0$; if A is not free then $\text{syz}\, A$ is a nontrivial module, and it can in turn be represented by a similar exact sequence; so writing

$\text{syz}^2 A = \text{syz}(\text{syz}\, A)$ for short, there is an exact sequence of ${}_V\mathcal{O}$-modules

$$0 \longrightarrow \text{syz}^2 A \longrightarrow {}_V\mathcal{O}^{r_1} \xrightarrow{\sigma_1} \text{syz}\, A \longrightarrow 0 .$$

If $\text{syz}\, A$ is not free so that $\text{syz}^2 A$ is also nontrivial, the construction can be repeated to yield yet another exact sequence of ${}_V\mathcal{O}$-modules

$$0 \longrightarrow \text{syz}^3 A \longrightarrow {}_V\mathcal{O}^{r_2} \xrightarrow{\sigma_2} \text{syz}^2 A \longrightarrow 0 ,$$

and so on. These sequences can be combined in a long exact sequence of ${}_V\mathcal{O}$-modules

$$\cdots \xrightarrow{\sigma_3} {}_V\mathcal{O}^{r_2} \xrightarrow{\sigma_2} {}_V\mathcal{O}^{r_1} \xrightarrow{\sigma_1} {}_V\mathcal{O}^{r} \xrightarrow{\sigma} A \longrightarrow 0$$

called the <u>minimal free resolution</u> (or minimal free homological resolution) of the ${}_V\mathcal{O}$-module A; and in this sequence $\text{syz}^j A = \text{image}\ \sigma_j = \text{kernel}\ \sigma_{j-1}$.

<u>Corollary 2 to Theorem 16</u>. For any exact sequence of ${}_V\mathcal{O}$-modules of the form

$$\cdots \xrightarrow{\tau_3} {}_V\mathcal{O}^{n_2} \xrightarrow{\tau_2} {}_V\mathcal{O}^{n_1} \xrightarrow{\tau_1} {}_V\mathcal{O}^{n} \xrightarrow{\tau} A \longrightarrow 0$$

there are isomorphisms

$$\text{image}\ \tau_j = \text{kernel}\ \tau_{j-1} \cong \text{syz}^j A \oplus {}_V\mathcal{O}^{m_j}$$

for some integers m_j, for $j = 1,2,3,\ldots$.

Proof. It follows from Theorem 16 that there is an isomorhpism $\theta: {}_V\mathcal{O}^n \longrightarrow {}_V\mathcal{O}^r \oplus {}_V\mathcal{O}^{n-r}$ where $r = \dim_{\mathbb{C}} A/{}_V\mathfrak{m}A$ such that in the modified exact sequence

$$\cdots \xrightarrow{\tau_3} {}_V\mathcal{O}^{n_2} \xrightarrow{\tau_2} {}_V\mathcal{O}^{n_1} \xrightarrow{\theta\tau_1} {}_V\mathcal{O}^r \oplus {}_V\mathcal{O}^{n-r} \xrightarrow{\tau\theta^{-1}} A \longrightarrow 0$$

necessarily $\tau\theta^{-1}(F,G) = \sigma(F)$ for $F \in {}_V\mathcal{O}^r$, $G \in {}_V\mathcal{O}^{n-r}$; thus the end of this exact sequence can be split off to yield the exact sequence

$$\cdots \xrightarrow{\tau_3} {}_V\mathcal{O}^{n_2} \xrightarrow{\tau_2} {}_V\mathcal{O}^{n_1} \xrightarrow{\theta\tau_1} \text{syz } A \oplus {}_V\mathcal{O}^{n-r} \longrightarrow 0 .$$

This shows in particular that image $\tau_1 \cong \text{syz } A \oplus {}_V\mathcal{O}^{m_1}$. Then since $\text{syz }(\text{syz } A \oplus {}_V\mathcal{O}^{m_1}) \cong \text{syz}^2 A$ as a consequence of Corollary 1 to Theorem 16, the desired corollary follows directly by a repetition of the preceding argument.

If $\text{syz}^j A = 0$ for some indices j then the smallest integer d such that $\text{syz}^{d+1} A = 0$ will be called the <u>homological dimension</u> of the ${}_V\mathcal{O}$-module A and will be denoted by $\text{hom dim}_{{}_V\mathcal{O}} A$ or more conveniently by $\text{hom dim}_V A$; and that none of the modules $\text{syz}^j A$ are trivial will be indicated by writing $\text{hom dim}_V A = \infty$. Thus the ${}_V\mathcal{O}$-module A is free precisely when $\text{hom dim}_V A = 0$. More generally, if $\text{hom dim}_V A = d < \infty$ then the minimal free resolution of A reduces to the exact sequence of ${}_V\mathcal{O}$-modules

$$0 \longrightarrow {}_V\mathcal{O}^{r_d} \xrightarrow{\sigma_d} {}_V\mathcal{O}^{r_{d-1}} \xrightarrow{\sigma_{d-1}} \cdots \xrightarrow{\sigma_2} {}_V\mathcal{O}^{r_1} \xrightarrow{\sigma_1} {}_V\mathcal{O}^{r} \xrightarrow{\sigma} A \longrightarrow 0$$

in which none of the kernels of the homomorphisms $\sigma, \sigma_1, \ldots, \sigma_{d-1}$ are free; and for any free resolution

$$\cdots \xrightarrow{\tau_3} {}_V\mathcal{O}^{n_2} \xrightarrow{\tau_2} {}_V\mathcal{O}^{n_1} \xrightarrow{\tau_1} {}_V\mathcal{O}^{n} \xrightarrow{\tau} A \longrightarrow 0 ,$$

the kernel of τ_j is a free ${}_V\mathcal{O}$-module whenever $j \geq d - 1$, but is not a free ${}_V\mathcal{O}$-module whenever $j \leq d - 2$.

Before turning to a discussion of the analytic significance of these concepts it is interesting to see them in a semi-local form as well, that is to say, in the context of analytic sheaves. If $\mathcal{A}$ is a coherent analytic sheaf over a complex analytic variety V then in an open neighborhood U of any point $0 \in V$ there is an exact sequence of analytic sheaves of the form

$${}_V\mathcal{O}^{r_1} \xrightarrow{\sigma_1} {}_V\mathcal{O}^{r} \xrightarrow{\sigma} \mathcal{A} \longrightarrow 0 ;$$

and since the kernel of σ_1 is also a coherent analytic sheaf then possibly after restricting the neighborhood U the exact sheaf sequence can be extended further to the left, and the process can obviously be continued. Thus in a sufficiently small open neighborhood U of the point $0 \in V$ there is an exact sequence of analytic sheaves of the form

$$(4) \quad {}_V\mathcal{O}^{r_d} \xrightarrow{\sigma_d} {}_V\mathcal{O}^{r_{d-1}} \xrightarrow{\sigma_{d-1}} \cdots \xrightarrow{\sigma_2} {}_V\mathcal{O}^{r_1} \xrightarrow{\sigma_1} {}_V\mathcal{O}^{r} \xrightarrow{\sigma} \mathcal{A} \longrightarrow 0$$

for any fixed integer d. Considering just the stalks over the point $0 \in V$ there results a free resolution of the ${}_V\mathcal{O}_0$-module $\mathcal{A}_0$; indeed it can be assumed that this is the minimal free resolution of the ${}_V\mathcal{O}_0$-module $\mathcal{A}_0$, since it is quite obvious that if $\mathcal{A}$, $\mathcal{B}$ are coherent analytic sheaves with $\mathcal{B} \subseteq \mathcal{A}$ and if $\mathcal{B}_0 = \mathcal{A}_0$ then the sheaves $\mathcal{A}$ and $\mathcal{B}$ coincide in a full open neighborhood of the point 0. On the one hand then, if $\text{hom dim}_{{}_V\mathcal{O}_0} \mathcal{A}_0 = d$ there is an exact sequence of sheaves of the above form where the stalk at $0 \in V$ of the kernel of σ_d is trivial, hence where the sheaf homomorphism σ_d is injective in an open neighborhood of the point $0 \in V$; and consequently $\text{hom dim}_{{}_V\mathcal{O}_p} \mathcal{A}_p \leq d$ at all points p of that neighborhood. Equivalently of course, for any coherent analytic sheaf $\mathcal{A}$ and any integer d the set $\{p \in V \mid \text{hom dim}_{{}_V\mathcal{O}_p} \mathcal{A}_p \leq d\}$ is an open subset of the complex analytic variety V, possibly the empty set though. On the other hand the following even more precise result can easily be established.

Corollary 3 to Theorem 16. For any coherent analytic sheaf $\mathcal{A}$ over a complex analytic variety V and any integer $d > 0$ the subset $\{p \in V \mid \text{hom dim}_{{}_V\mathcal{O}_p} \mathcal{A}_p \geq d\}$ is a proper complex analytic subvariety of V.

Proof. It is clear from the definition that the set of points $p \in V$ at which $\text{hom dim}_{{}_V\mathcal{O}_p} \mathcal{A}_p \leq d-1$ is precisely the set of those points p at which $\text{syz}^d \mathcal{A}_p = 0$, or equivalently at which $\operatorname{syz}^{d-1} \mathcal{A}_p$ is a free ${}_V\mathcal{O}_p$-module, where $\operatorname{syz}^0 \mathcal{A}_p = \mathcal{A}_p$. Consider an exact sequence of the form (4) over an open neighborhood U of some point of V; and let $\mathcal{S} \subseteq {}_V\mathcal{O}^{r_{d-2}}$ be the image of the sheaf homomorphism σ_{d-1}, so that there is an exact sequence of analytic sheaves

$$ {}_V\mathcal{O}^{r_d} \xrightarrow{\sigma_d} {}_V\mathcal{O}^{r_{d-1}} \xrightarrow{\sigma_{d-1}} \mathcal{S} \longrightarrow 0 $$

over the neighborhood U. It follows Corollary 2 to Theorem 16 that at any point $p \in U$ the stalk $\mathcal{S}_p \cong \operatorname{syz}^{d-1} \mathcal{A}_p \oplus {}_V\mathcal{O}_p^m$ for some m, and as noted in the proof of Corollary 2 to Theorem 15 a direct summand of a free ${}_V\mathcal{O}_p$-module is also free; it is then clear that $\operatorname{syz}^{d-1} \mathcal{A}_p$ is a free ${}_V\mathcal{O}_p$-module precisely when $\mathcal{S}_p$ is a free ${}_V\mathcal{O}_p$-module. Now the sheaf homomorphism σ_d is described by a matrix H of functions holomorphic in U, and it is evident that the set of those points $p \in U$ at which $\operatorname{rank} H(p) < \max_{q \in U} \operatorname{rank} H(q)$ is a proper complex analytic subvariety of the neighborhood U, possibly the empty set of course; hence to conclude the proof it is enough just to show that $\mathcal{S}_p$ is a free ${}_V\mathcal{O}_p$-module precisely when $\operatorname{rank} H(p) = \max_{q \in U} \operatorname{rank} H(q)$ for a connected open neighborhood U. On the one hand suppose that $\operatorname{rank} H(p) = \max_{q \in U} \operatorname{rank} H(q) = n$ for some point $p \in U$. After a suitable automorphism of the free sheaves ${}_V\mathcal{O}^{r_d}$, ${}_V\mathcal{O}^{r_{d-1}}$ it can be assumed that

$$H(p) = \begin{pmatrix} H_1(p) & 0 \\ 0 & 0 \end{pmatrix}$$

where $H_1(p)$ is a nonsingular matrix of rank n; but then $H = \begin{pmatrix} H_1 & 0 \\ 0 & 0 \end{pmatrix}$ where H_1 is an $n \times n$ square matrix of holomorphic functions in U and $H_1(q)$ is nonsingular for all points q sufficiently near p, so at p the image of σ_d is a direct summand ${}_V\mathcal{O}_p^n \subseteq {}_V\mathcal{O}_p^{r_{d-1}}$ and consequently $\mathcal{S}_p$ is a free ${}_V\mathcal{O}_p$-module. On the other hand if $\mathcal{S}_p$ is a free ${}_V\mathcal{O}_p$-module of rank m it follows from Theorem 16 that there is an isomorphism θ: ${}_V\mathcal{O}_p^{r_{d-1}} \longrightarrow {}_V\mathcal{O}_p^m \oplus {}_V\mathcal{O}_p^n$ such that $\theta\sigma_d$ is a surjective homomorphism from ${}_V\mathcal{O}_p^{r_d}$ onto ${}_V\mathcal{O}_p^n$. This homomorphism is represented by the matrix of holomorphic functions $GH = \begin{pmatrix} 0 & 0 \\ H_2 & H_1 \end{pmatrix}$, where G is nonsingular near p and H_1 is nonsingular of rank n near p; and since U is connected it is evident that $\text{rank}\, H(p) = \text{rank}\, H(q)$ for all q near p, hence that $\text{rank}\, H(p) = \max_{q \in U} \text{rank}\, H(q)$. That suffices to conclude the proof of the corollary.

In particular note that an arbitrary coherent analytic sheaf over a complex analytic variety is locally free outside a proper complex analytic subvariety.

(d) For any germ V of a complex analytic variety a finite analytic mapping φ: $V \to \mathbb{C}^k$ exhibits the local ring ${}_V\mathcal{O}$ as a finitely generated ${}_k\mathcal{O}$-module, the homological dimension of which will be denoted by hom $\dim_\varphi V$; the minimal value of hom $\dim_\varphi V$ for all finite analytic mappings $\varphi\colon V \to \mathbb{C}^k$ where $k = \dim V$ will be called simply the <u>homological dimension of the germ</u> V and will be denoted by hom dim V. Perfect germs of complex analytic varieties can thus be characterized as those germs V for which hom dim $V = 0$, and in general hom dim V can be viewed as a measure of the extent to which a germ V fails to be perfect. This measure is particularly convenient in discussing some properties of general complex analytic varieties analogous to the analytic continuation properties of perfect varieties described in Theorem 15. The reader should perhaps be warned that in this discussion it is necessary to invoke more cohomological machinery than has been so far required in these notes.

<u>Theorem 17</u>. If V is a germ of a complex analytic variety with hom dim $V = d$ then any complex analytic subvariety $W \subset V$ with $\dim W \leq \dim V - d - 2$ is a removable singularity for holomorphic functions.

Proof. If hom dim $V = d$ and $\dim V = k$ then there is a finite analytic mapping $\varphi\colon V \to \mathbb{C}^k$ exhibiting ${}_V\mathcal{O}$ as a finitely generated ${}_k\mathcal{O}$-module of homological dimension d; when considered as an ${}_k\mathcal{O}$-module ${}_V\mathcal{O}$ can be viewed as the stalk at the origin of the direct image sheaf $\varphi_*({}_V\mathcal{O})$, and consequently that sheaf admits a free resolution of the form

$$0 \to {}_k\mathcal{O}^{r_d} \xrightarrow{\sigma_d} {}_k\mathcal{O}^{r_{d-1}} \xrightarrow{\sigma_{d-1}} \cdots \xrightarrow{\sigma_2} {}_k\mathcal{O}^{r_1} \xrightarrow{\sigma_1} {}_k\mathcal{O}^{r} \xrightarrow{\sigma} \varphi_*({}_V\mathcal{O}) \to 0$$

over some open neighborhood U of the origin in $\mathbb{C}^k$. This exact sequence can of course be rewritten as a set of short exact sequences of the form

$$0 \longrightarrow {}_k\mathcal{O}^{r_d} \xrightarrow{\sigma_d} {}_k\mathcal{O}^{r_{d-1}} \xrightarrow{\sigma_{d-1}} \mathcal{S}_{d-1} \longrightarrow 0$$

$$0 \longrightarrow \mathcal{S}_{d-1} \xrightarrow{i} {}_k\mathcal{O}^{r_{d-2}} \xrightarrow{\sigma_{d-2}} \mathcal{S}_{d-2} \longrightarrow 0$$

$$\cdots$$

$$0 \longrightarrow \mathcal{S}_{2} \xrightarrow{i} {}_k\mathcal{O}^{r_1} \xrightarrow{\sigma_1} \mathcal{S}_{1} \longrightarrow 0$$

$$0 \longrightarrow \mathcal{S}_{1} \xrightarrow{i} {}_k\mathcal{O}^{r} \xrightarrow{\sigma} \varphi_*({}_V\mathcal{O}) \longrightarrow 0 ,$$

where the coherent analytic sheaf $\mathcal{S}_j$ is the image of the sheaf homomorphism σ_j and i denotes the inclusion mapping. Now for any complex analytic subvariety $W \subset V$ the image $W_0 = \varphi(W)$ is a complex analytic subvariety of the open subset U in $\mathbb{C}^k$; and if W_0 is a proper subvariety of U then the complementary set $U - W_0$ is nonempty, and over that set the exact cohomology sequences associated to the above short exact sheaf sequences contain the segments

$$\cdot \to H^{d-1}(U-W_0, {}_k\mathcal{O}^{r_{d-1}}) \xrightarrow{\sigma^*_{d-1}} H^{d-1}(U-W_0, \mathcal{S}_{d-1}) \xrightarrow{\delta^*} H^d(U-W_0, {}_k\mathcal{O}^{r_d}) \longrightarrow \cdot$$

$$\cdot \to H^{d-2}(U-W_0, {}_k\mathcal{O}^{r_{d-2}}) \xrightarrow{\sigma^*_{d-2}} H^{d-2}(U-W_0, \mathcal{S}_{d-2}) \xrightarrow{\delta^*} H^{d-1}(U-W_0, \mathcal{S}_{d-1}) \to \cdot$$

$$\cdots$$

$$\cdot \longrightarrow H^1(U-W_0, {}_k\mathcal{O}^{r_1}) \xrightarrow{\sigma_1^*} H^1(U-W_0, \mathcal{S}_1) \xrightarrow{\delta^*} H^2(U-W_0, \mathcal{S}_2) \longrightarrow \cdot$$

$$\cdot \longrightarrow \Gamma(U-W_0, {}_k\mathcal{O}^r) \xrightarrow{\sigma^*} \Gamma(U-W_0, \varphi_*({}_V\mathcal{O})) \xrightarrow{\delta^*} H^1(U-W_0, \mathcal{S}_1) \longrightarrow \cdot$$

Note that if $\dim W \leq \dim V - d - 2$ then $\dim W_0 = \dim W \leq k - d - 2$ It is then a special case of a theorem of Frenkel that for a subvariety W_0 with this dimensional restriction the neighborhood U can be so chosen that $H^p(U-W_0, {}_k\mathcal{O}) = 0$ for $1 \leq p \leq d$; this assertion is perhaps not in the complex analyst's standard cohomological repertoire, so a proof is included separately in the appendix to these notes, (Corollary 1 to Theorem 22.) Applying this result to the above segments of exact cohomology sequences, it follow consecutively that $0 \cong H^{d-1}(U-W_0, \mathcal{S}_{d-1}) \cong H^{d-2}(U-W_0, \mathcal{S}_{d-2}) \cong \cdots \cong H^2(U-W_0, \mathcal{S}_2) \cong H^1(U-W_0, \mathcal{S}_1)$, and consequently that the homomorphism

$$\sigma^*: \ \Gamma(U-W_0, {}_k\mathcal{O}^r) \longrightarrow \Gamma(U-W_0, \varphi_*({}_V\mathcal{O}))$$

is surjective; the cases $d = 0,1$ are slightly special but only rather trivially so, and the modifications necessary in the preceding argument in these cases will be left to the reader, the conclusion being that in these cases as well the homomorphism σ^*

is surjective. The restriction to $V - \varphi^{-1}(W_0) \subseteq V - W$ of any holomorphic function f on $V - W$ can be viewed as a section $f \in \Gamma(U-W_0, \varphi_*({}_V\mathcal{O}))$; and there thus exists a section $F \in \Gamma(U-W_0, {}_k\mathcal{O}^r)$ such that $\sigma^*F = f$. However F is merely a set of r holomorphic functions on $U - W_0$, and since $\dim W_0 \leq k - d - 2 \leq k - 2$ it follows from the extended Riemann removable singularities theorem that F extends to a section $\tilde{F} \in \Gamma(U, {}_k\mathcal{O}^r)$; and the image $\sigma^*\tilde{F} = \tilde{f} \in \Gamma(U, \varphi_*({}_V\mathcal{O}))$ can be viewed as a holomorphic function on V such that $\tilde{f}|(V - \varphi^{-1}(W_0)) = f|(V - \varphi^{-1}(W_0))$. For any irreducible component V_1 of the germ V the function $\tilde{f}$ is then holomorphic on all of V_1, the function f is holomorphic on $V_1 - V_1 \cap W$, and these two functions agree on $V_1 - V_1 \cap \varphi^{-1}(W_0)$ where of course $V_1 - V_1 \cap \varphi^{-1}(W_0) \subseteq V_1 - V_1 \cap W$. If either $V_1 \cap \varphi^{-1}(W_0)$ is a proper analytic subvariety of V_1 or $V_1 \cap \varphi^{-1}(W_0) = V_1 \cap W$ then the functions f and $\tilde{f}$ agree on all of $V_1 - V_1 \cap W$, but if $V_1 = V_1 \cap \varphi^{-1}(W_0)$ and $V_1 \cap W \subset V_1 \cap \varphi^{-1}(W_0)$ then these two functions need not agree on $V_1 - V_1 \cap W$. That is at least enough to prove the theorem for all cases except those in which the germ V has an irreducible component V_1 and the germ W is such that $V_1 \not\subseteq W$ but $\varphi(V_1) \subseteq \varphi(W)$ for all finite analytic mappings $\varphi: V \to \mathbb{C}^k$ exhibiting ${}_V\mathcal{O}$ as an ${}_k\mathcal{O}$-module of homological dimension d. It is easy to see though that this exceptional situation cannot occur. For if there were such subvarieties V_1 and W in V then letting V_2 be the union of all the irreducible components of V except for V_1 and setting $X = V_1 \cap V_2$ it would follow that

$\dim X < \dim V_1 \leq \dim W \leq \dim V - d - 2$, and the part of the theorem already proved would apply to show that $\varphi^{-1}(\varphi(X))$ is a removable singularity for holomorphic functions on V; but that is clearly impossible, as is evident upon considering the function which is zero on $V_1 - V_1 \cap \varphi^{-1}(\varphi(X))$ and one on $V_2 - V_2 \cap \varphi^{-1}(\varphi(X))$. That suffices to conclude the proof of the entire theorem.

It was noted earlier that perfect germs of complex analytic varieties are necessarily pure-dimensional; for a general germ of complex analytic variety the homological dimension bounds the extent to which that germ fails to be pure-dimensional, in the following sense.

Corollary 1 to Theorem 17. If V is a germ of a complex analytic variety with $\text{hom dim } V = d$ then for any irreducible component V_1 of the germ V

$$\dim V - d \leq \dim V_1 \leq \dim V .$$

Proof. If there were an irreducible component V_1 of the germ V with $\dim V_1 \leq \dim V - d - 1$ then letting V_2 be the union of the other irreducible components of V and setting $W = V_1 \cap V_2$ it would follow that $\dim W \leq \dim V_1 - 1 \leq \dim V - d - 2$, and hence by Theorem 17 the subset W would be a removable singularity for holomorphic functions on V; but that is clearly impossible, as is evident upon considering the holomorphic function on $V - W$ which is zero on $V_1 - W$ and one on $V_2 - W$.

That contradiction suffices to conclude the proof of the corollary.

Since the weakly holomorphic functions on a germ V of a complex analytic variety are necessarily holomorphic on $V - \mathcal{S}(V)$ another immediate consequence of Theorem 17 is the following.

<u>Corollary 2 to Theorem 17</u>. If V is a germ of a complex analytic variety with $\text{hom dim } V = d$ and if $\dim \mathcal{S}(V) \leq \dim V - d - 2$ then V is normal.

Using these corollaries it is quite easy to construct examples of germs of complex analytic varieties with relatively large homological dimension. For example if V is a nonnormal germ of a complex analytic variety and if $\dim V - \dim \mathcal{S}(V) = r$ then by Corollary 2 to Theorem 17 necessarily $\text{hom dim } V \geq r - 1$; in particular if V is nonnormal but has an isolated singularity then $\text{hom dim } V \geq \dim V - 1$. It will later be demonstrated that $\text{hom dim } V \leq \dim V - 1$ for arbitrary germs V of complex analytic varieties, and the example of a nonnormal germ with an isolated singularity shows that this maximal value for the homological dimension of a germ V is actually attained. Examples of normal germs having relatively large homological dimension are apparently rather harder to come by.

Turning from germs of varieties to varieties themselves, it is natural to say that a complex analytic variety V is of homological dimension d at a point $p \in V$ if the germ of V at the point p is of homological dimension d; the homological dimension of the variety V at a point $p \in V$ will be denoted by hom dim V_p. If hom dim $V_0 = d$ at some point $0 \in V$ then there is a finite analytic mapping $\varphi\colon V \to \mathbb{C}^k$ in an open neighborhood of 0, taking $0 \in V$ to the origin $0 \in \mathbb{C}^k$ and exhibiting ${}_V\mathcal{O}_0$ as an ${}_k\mathcal{O}_0$-module of homological dimension d; hence as before, since ${}_V\mathcal{O}_0$ can be viewed as the stalk at $0 \in \mathbb{C}^k$ of the direct image sheaf $\varphi_*({}_V\mathcal{O})$, there is an exact sequence of analytic sheaves of the form

$$0 \to {}_k\mathcal{O}^{r_d} \xrightarrow{\sigma_d} {}_k\mathcal{O}^{r_{d-1}} \xrightarrow{\sigma_{d-1}} \cdots \xrightarrow{\sigma_2} {}_k\mathcal{O}^{r_1} \xrightarrow{\sigma_1} {}_k\mathcal{O}^{r} \xrightarrow{\sigma} \varphi_*({}_V\mathcal{O}) \to 0$$

in some open neighborhood of the origin in $\mathbb{C}^k$. Then since $\varphi_*({}_V\mathcal{O})_{\varphi(p)} = {}_V\mathcal{O}_p \oplus {}_V\mathcal{O}_{p_1} \oplus \cdots \oplus {}_V\mathcal{O}_{p_n}$ for any point $p \in V$ sufficiently near 0, where $\varphi^{-1}(p) = \{p_1,\ldots,p_n\}$, it follows immediately from Corollaries 1 and 2 to Theorem 16 that

$$\begin{aligned} 0 &\cong \operatorname{syz}^{d+1}({}_V\mathcal{O}_p \oplus {}_V\mathcal{O}_{p_1} \oplus \cdots \oplus {}_V\mathcal{O}_{p_n}) \\ &\cong \operatorname{syz}^{d+1}\,{}_V\mathcal{O}_p \oplus \cdots \oplus \operatorname{syz}^{d+1}\,{}_V\mathcal{O}_{p_n}\,. \end{aligned}$$

and hence that $\operatorname{syz}^{d+1}\,{}_V\mathcal{O}_p = 0$; consequently hom dim $V_p \le d$ for all points $p \in V$ sufficiently near 0. That is to say, for any integer d the set $\{p \in V \mid \text{hom dim } V_p \le d\}$ is an open subset of the complex analytic variety V. The anticipated more precise result, that for any integer d the set $\{p \in V \mid \text{hom dim } V_p \ge d\}$ is a complex analytic subvariety of V, is also true; but it is more convenient to postpone the proof of that assertion.

(e) Although perfect germs of complex analytic varieties need not be irreducible, it was observed earlier in these notes that their local rings contain a considerable number of elements which are not divisors of zero; indeed if $\varphi: V \to \mathbb{C}^k$ is a finite analytic mapping exhibiting the local ring ${}_V\mathcal{O}$ of the germ V of complex analytic variety as a free ${}_k\mathcal{O}$-module then the images in ${}_V\mathcal{O}$ of the coordinate functions $z_1, \ldots, z_k$ in $\mathbb{C}^k$ are relatively independent elements of ${}_V\mathcal{O}$ which are not divisors of zero. This observation can be made more precise, and leads to another interesting and useful interpretation of the homological dimension of a germ of complex analytic variety; actually in the more purely algebraic treatment of local rings it is this interpretation rather than the definition used here that plays the primary role. To begin the discussion it may be useful to review some properties of zero-divisors in a slightly more general situation.

Suppose then that A is a module over the local ring ${}_V\mathcal{O}$ of some germ V of a complex analytic variety. For any subset $S \subseteq A$ the <u>annihilator</u> of S is defined to be the subset of the ring ${}_V\mathcal{O}$ consisting of those elements $f \in {}_V\mathcal{O}$ such that $f \cdot s = 0$ for all $s \in S$, and is denoted by $\operatorname{ann} S$; thus

$$\operatorname{ann} S = \{f \in {}_V\mathcal{O} \mid f \cdot S = 0\} .$$

It is evident that the annihilator of any subset of A is an ideal in the ring ${}_V\mathcal{O}$. In particular to any nonzero element $a \in A$

there is associated the ideal $\operatorname{ann} a \subseteq {}_V\mathcal{O}$; and the union

$$\bigcup_{\{a\in A \mid a\neq 0\}} \operatorname{ann} a$$

is precisely the set of zero-divisors for the module A. The ideals in ${}_V\mathcal{O}$ of the form $\operatorname{ann} a$ where $a \neq 0$ can be characterized as those ideals $\mathfrak{A} \subseteq {}_V\mathcal{O}$ such that the module A contains a submodule isomorphic to ${}_V\mathcal{O}/\mathfrak{A}$; for if $\mathfrak{A} = \operatorname{ann} a$ then ${}_V\mathcal{O}\cdot a$ is a submodule of A isomorphic to ${}_V\mathcal{O}/\mathfrak{A}$, and if $B \subseteq A$ is a submodule for which there exists an isomorphism $\sigma\colon {}_V\mathcal{O}/\mathfrak{A} \longrightarrow B$ for some ideal $\mathfrak{A}$ then $\sigma(1) = a \in B$ is a nonzero element such that $\mathfrak{A} = \operatorname{ann} a$. Note that the maximal elements among the set of ideals $\{\operatorname{ann} a\}$ must actually be prime ideals. To see this, if $\operatorname{ann} a$ is a maximal element among this set of ideals (in the sense that $\operatorname{ann} a \subseteq \operatorname{ann} b$ for any nonzero element $b \in A$ implies that $\operatorname{ann} a = \operatorname{ann} b$), then whenever $fg \in \operatorname{ann} a$ but $f \notin \operatorname{ann} a$ necessarily $fg\cdot a = 0$ but $f\cdot a \neq 0$, hence $g \in \operatorname{ann} f\cdot a$; but clearly $\operatorname{ann} a \subseteq \operatorname{ann} f\cdot a$, so that from maximality it follows that $\operatorname{ann} a = \operatorname{ann} f\cdot a$ and hence that $g \in \operatorname{ann} a$, so that $\operatorname{ann} a$ is a prime ideal. The maximal elements among the set of ideals $\operatorname{ann} a$ for $a \neq 0$, or equivalently the proper prime ideals in ${}_V\mathcal{O}$ of the form $\operatorname{ann} a$, will be called the associated prime ideals for the module A; and the set of all these associated prime ideals will be denoted by $\operatorname{ass} A$. Thus the set of zero-divisors for the module A can be described equivalently as the union of the associated prime ideals for the module A, that is to say as the set $\bigcup_{\mathfrak{P} \in \text{ass}\, A} \mathfrak{P}$.

For any exact sequence of ${}_V\mathcal{O}$-modules of the form $0 \to A' \to A \to A'' \to 0$ it is quite easy to see that $\text{ass}\, A \subseteq \text{ass}\, A' \cup \text{ass}\, A''$. Indeed suppose that $\mathfrak{P}$ is a proper prime ideal in ${}_V\mathcal{O}$ such that $\mathfrak{P} \in \text{ass}\, A$; there is then a submodule $B \subseteq A$ isomorphic to ${}_V\mathcal{O}/\mathfrak{P}$. If $B \cap A' = 0$ then the image of B in A'' is a submodule of A'' isomorphic to B and hence to ${}_V\mathcal{O}/\mathfrak{P}$, and consequently $\mathfrak{P} \in \text{ass}\, A''$. On the other hand if there is a nonzero element $b \in B \cap A'$, then since $B \cong {}_V\mathcal{O}/\mathfrak{P}$ is an integral domain, for any element $f \in {}_V\mathcal{O}$ it follows that $f \cdot b = 0$ precisely when $f \in \mathfrak{P}$; hence $\mathfrak{P} = \text{ann}\, b$, and consequently $\mathfrak{P} \in \text{ass}\, A'$.

It is in turn a simple consequence of this last observation that for a finitely generated ${}_V\mathcal{O}$-module A the set $\text{ass}\, A$ is a finite set of prime ideals. For if $A \neq 0$ and $\mathfrak{P}_1 \in \text{ass}\, A$ there is a submodule $A_1 \subseteq A$ such that $A_1 \cong {}_V\mathcal{O}/\mathfrak{P}_1$, and clearly $\text{ass}\, A_1 = \text{ass}\, {}_V\mathcal{O}/\mathfrak{P}_1 = \{\mathfrak{P}_1\}$; if $A/A_1 \neq 0$ and $\mathfrak{P}_2 \in \text{ass}\, A/A_1$ there is a submodule $A_2/A_1 \subseteq A/A_1$ such that $A_2/A_1 \cong {}_V\mathcal{O}/\mathfrak{P}_2$, and $\text{ass}\, A_2/A_1 = \{\mathfrak{P}_2\}$; and if $A/A_2 \neq 0$ the process can be repeated. There thus results a chain of submodules $A_1 \subset A_2 \subset A_3 \subset \ldots$ of A such that $\text{ass}\, A_1 = \{\mathfrak{P}_1\}$ and $\text{ass}\, A_i/A_{i-1} = \{\mathfrak{P}_i\}$ for $i > 1$; and since A is finitely generated this ascending chain of submodules must eventually terminate, so that $A_n = A$ for some index n. Then applying the preceding observation inductively it follows that

$$\text{ass } A = \text{ass } A_n \subseteq \text{ass } (A_n/A_{n-1}) \cup \text{ass } A_{n-1}$$

$$\subseteq \text{ass } (A_n/A_{n-1}) \cup \text{ass } (A_{n-1}/A_{n-2}) \cup \ldots \cup \text{ass } (A_2/A_1) \cup \text{ass } A_1$$

$$\subseteq \{\mathfrak{p}_n, \mathfrak{p}_{n-1}, \ldots, \mathfrak{p}_2, \mathfrak{p}_1\}$$

hence ass A is a finite set of prime ideals as desired. It follows from this that the set of zero-divisors for a finitely generated ${}_V\mathcal{O}$-module is the union of finitely many proper prime ideals of ${}_V\mathcal{O}$.

Now for any finitely generated ${}_V\mathcal{O}$-module A a sequence of elements $\{f_1,\ldots,f_r\}$ where $f_i \in {}_V\mathfrak{m}$ will be called an <u>A-sequence of length r</u> if f_i is not a zero-divisor for the ${}_V\mathcal{O}$-module $A/(f_1 \cdot A + \ldots + f_r \cdot A)$ for $1 \leq i \leq r$; thus f_1 is not a zero-divisor for A, f_2 is not a zero-divisor for $A/f_1 \cdot A$, and so on. For any A-sequence $\{f_1,\ldots,f_r\}$ either there exists an element $f_{r+1} \in {}_V\mathfrak{m}$ which is not a zero-divisor for $A/(f_1 \cdot A + \ldots + f_r \cdot A)$ or all elements of ${}_V\mathfrak{m}$ are zero-divisors for $A/(f_1 \cdot A + \ldots + f_r \cdot A)$; in the first case $\{f_1,\ldots,f_r,f_{r+1}\}$ is also an A-sequence, providing an extension of the initial A-sequence, while in the second case $\{f_1,\ldots,f_r\}$ is a <u>maximal A-sequence</u> in the sense that it cannot be extended to an A-sequence of greater length. If $\{f_1,f_2,\ldots\}$ is an A-sequence and $A_i = f_1 \cdot A + \ldots + f_i \cdot A \subset A$ then whenever $a \in A - A_i$ necessarily $f_{i+1} \cdot a \in A_{i+1} - A_i$, so that $A_i \subset A_{i+1}$; the submodules A_i thus form a strictly increasing chain of submodules of A, and since A is finitely generated this chain must necessarily be finite.

Therefore every A-sequence can be extended to a maximal A-sequence. The maximum of the set of integers r such that there exists an A-sequence of length r will be called the <u>profundity</u> of the ${}_V\mathcal{O}$-module A, and will be denoted by $\text{prof}_{{}_V\mathcal{O}} A$ or more conveniently just by $\text{prof}_V A$. (The French word profondeur is commonly used here; the English word profundity seems more natural and convenient than either depth or grade, which are also sometimes used.) If the profundity of the ${}_V\mathcal{O}$-module is finite then all maximal A-sequences have bounded lengths; actually a great deal more can be asserted.

<u>Theorem 18</u>. Let A be a finitely generated ${}_V\mathcal{O}$-module for some germ V of a complex analytic variety. If $\{f_1,\ldots,f_r\}$ is an A-sequence then any permutation of this sequence is also an A-sequence. All maximal A-sequences are of the same length, and this common length is of course the profundity of A; consequently $0 \leq \text{prof}_V A < \infty$.

Proof. Note that $\{f_1,\ldots,f_r\}$ is an A-sequence if and only if $\{f_1,\ldots,f_s\}$ is an A-sequence and $\{f_{s+1},\ldots,f_r\}$ is an $(A/f_1 \cdot A + \ldots + f_s \cdot A)$-sequence, for any $s \leq r$; and since any permutation can be built up from transpositions then in order to show that any permutation of an A-sequence is also an A-sequence it suffices to show that if $\{f_1,f_2\}$ is an A-sequence then so is $\{f_2,f_1\}$. That $\{f_1,f_2\}$ is an A-sequence is equivalent to the two conditions: (i) $f_1 \cdot a = 0$ for some $a \in A$ implies $a = 0$; (ii) $f_2 \cdot a = f_1 \cdot b$ for some $a,b \in A$ implies $a = f_1 \cdot b_1$ for some

$b_1 \in A$. Now if $f_2 \cdot a = 0$ for some $a \in A$ then from (ii) necessarily $a = f_1 \cdot b_1$ for some $b_1 \in A$, and $0 = f_2 \cdot a = f_1 f_2 \cdot b_1$ so from (i) also $f_2 \cdot b_1 = 0$; repeating this argument shows that $b_1 = f_1 \cdot b_2$ for some $b_2 \in A$ and $f_2 \cdot b_2 = 0$, and so on. Thus $a = f_1 \cdot b_1 = f_1^2 \cdot b_2 = \ldots$, so that $a \in {}_V\mathfrak{m}^n \cdot A$ for every integer n; and it then follows from Nakayama's lemma that $a = 0$. On the other hand if $f_1 \cdot a = f_2 \cdot b$ for some $a, b \in A$ then from (ii) necessarily $b = f_1 \cdot b_1$ for some $b_1 \in A$; but $f_1 \cdot a = f_2 \cdot b = f_1 f_2 \cdot b_1$, so from (i) then $a = f_2 \cdot b_1$. Therefore $\{f_2, f_1\}$ is also an A-sequence as desired.

It is convenient at this stage of the proof to consider separately the simplest special cases. First $\text{prof}_V A = 0$ means precisely that there are no A-sequences at all, or equivalently that all elements of ${}_V\mathfrak{m}$ are zero-divisors for A. In that case ${}_V\mathfrak{m} = \bigcup_{\mathfrak{p} \in \text{ass}\, A} \mathfrak{p}$, and since it is well known that an ideal which is the union of finitely many prime ideals must coincide with one of them, necessarily ${}_V\mathfrak{m} \in \text{ass}\, A$; hence ${}_V\mathfrak{m} = \text{ann}\, a$ for some nonzero element $a \in A$. Since the converse is quite obvious it follows that $\text{prof}_V A = 0$ if and only if ${}_V\mathfrak{m} \cdot a = 0$ for some nonzero element $a \in A$.

Next $\text{prof}_V A = 1$ means precisely that there are A-sequences, but all are of the form $\{f\}$; in particular if $\text{prof}_V A = 1$ then $\{f\}$ is a maximal A-sequence for every $f \in {}_V\mathfrak{m}$ which is not a zero-divisor for A. Note that in general an A-sequence $\{f\}$ is maximal if and only if $\text{prof}_V A/f \cdot A = 0$;

and as a consequence of the observation in the preceding paragraph, $\text{prof}_V\, A/f\cdot A = 0$ if and only if ${}_V\mathfrak{m}\cdot a \subseteq f\cdot A$ for some element $a \in A$ such that $a \notin f\cdot A$. To rephrase this condition, for any submodule $B \subseteq A$ let

$$[B:{}_V\mathfrak{m}]_A = \{a \in A \mid {}_V\mathfrak{m}\cdot a \subseteq B\} ,$$

noting that this is a submodule of A such that $B \subset [B:{}_V\mathfrak{m}]_A$; and with this notation an A-sequence $\{f\}$ is maximal if and only if $[f\cdot A:{}_V\mathfrak{m}]_A/f\cdot A \neq 0$. Now if f,g are two elements of ${}_V\mathfrak{m}$ then for any $a \in [f\cdot A:{}_V\mathfrak{m}]_A$ necessarily $g\cdot a \in {}_V\mathfrak{m}\cdot a \subseteq f\cdot A$; hence there must exist an element $\sigma(a) \in A$ such that $g\cdot a = f\cdot\sigma(a)$. If f is not a zero-divisor for the module A the element $\sigma(a)$ is uniquely determined by a, and the mapping $a \to \sigma(a)$ is then evidently a module homomorphism; since $f{}_V\mathfrak{m}\cdot\sigma(a) = g{}_V\mathfrak{m}\cdot a \subseteq fg\cdot A$ it also follows that ${}_V\mathfrak{m}\cdot\sigma(a) \subseteq g\cdot A$ hence that $\sigma(a) \in [g\cdot A:{}_V\mathfrak{m}]_A$, and in addition if $a = f\cdot b \in f\cdot A$ then $g\cdot a = fg\cdot b$ hence $\sigma(a) = g\cdot b \in g\cdot A$. Thus if f is not a zero-divisor for the module A then σ induces a module homomorphism

$$\sigma^*:\ [f\cdot A:{}_V\mathfrak{m}]_A/f\cdot A \longrightarrow [g\cdot A:{}_V\mathfrak{m}]_A/g\cdot A ;$$

and if g is also not a zero-divisor it is apparent by symmetry that the corresponding construction with f and g interchanged induces the homomorphism inverse to σ^*. Consequently if $\{f\}$ and $\{g\}$ are both A-sequences then

$$[f\cdot A :_V \mathfrak{m}]_A / f\cdot A \cong [g\cdot A :_V \mathfrak{m}]_A / g\cdot A ;$$

and therefore $\{f\}$ is a maximal A-sequence precisely when $\{g\}$ is a maximal A-sequence. In summary if $\text{prof}_V A = 1$ then $\{f\}$ is a maximal A-sequence for every $f \in {}_V\mathfrak{m}$ which is not a zero-divisor for A; and conversely if there exists a maximal A-sequence of the form $\{f\}$ then $\text{prof}_V A = 1$.

Returning to the general case again, suppose that $\{f_1,\ldots,f_r\}$ and $\{g_1,\ldots,g_s\}$ are two maximal A-sequences with $r \leq s$; to conclude the proof of the theorem it is only necessary to show that $r = s$. Note first that there must exist an element $f \in {}_V\mathfrak{m}$ such that $\{f_1,\ldots,f_{r-1},f\}$ and $\{g_1,\ldots,g_{s-1},f\}$ are still A-sequences. Indeed since f_r is not a zero divisor for $A' = A/(f_1\cdot A + \ldots + f_{r-1}\cdot A)$ it follows that ${}_V\mathfrak{m} \supset \bigcup_{\mathfrak{p}' \in \text{ass } A'} \mathfrak{p}'$, and since g_s is not a zero-divisor for $A'' = A/(g_1\cdot A + \ldots + g_{s-1}\cdot A)$ it follows that ${}_V\mathfrak{m} \supset \bigcup_{\mathfrak{p}'' \in \text{ass } A''} \mathfrak{p}''$; but then necessarily

$${}_V\mathfrak{m} \supset \Big(\bigcup_{\mathfrak{p}' \in \text{ass } A'} \mathfrak{p}'\Big) \cup \Big(\bigcup_{\mathfrak{p}'' \in \text{ass } A''} \mathfrak{p}''\Big) ,$$

hence there is an element $f \in {}_V\mathfrak{m}$ which is not a zero-divisor either for A' or for A'', as desired. Note next that $\{f_1,\ldots,f_{r-1},f\}$ and $\{g_1,\ldots,g_{s-1},f\}$ are still maximal A-sequences. Indeed since $\{f_1,\ldots,f_{r-1},f_r\}$ is a maximal A-sequence then $\{f_r\}$ is a maximal A'-sequence; but then as in the special case considered above $\text{prof}_V A' = 1$, hence $\{f\}$ is also a maximal A'-sequence and consequently $\{f_1,\ldots,f_{r-1},f\}$ is a maximal A-sequence as desired. Since any permutation of a maximal A-sequence is also clearly a maximal A-sequence, the preceding argument can be iterated to yield maximal A-sequences of the form $\{f_1,\ldots,f_r\}$ and $\{f_1,\ldots,f_r,g_{r+1},\ldots,g_s\}$. However since $\{f_1,\ldots,f_r\}$ is a maximal A-sequence then $\text{prof}_V\,(A/f_1\cdot A + \ldots + f_r\cdot A) = 0$; and since $\{g_{r+1},\ldots,g_s\}$ must be an $(A/f_1\cdot A + \ldots + f_r\cdot A)$-sequence necessarily $r = s$, and the proof of the theorem is thereby concluded.

One useful additional property of profundity is conveniently inserted here as part of the general discussion.

<u>Corollary 1 to Theorem 18</u>. For any exact sequence of ${}_V\mathcal{O}$-modules of the form

$$0 \longrightarrow A' \longrightarrow A \longrightarrow A'' \longrightarrow 0$$

it follows that $\text{prof}_V\, A \geq \min\,(\text{prof}_V\, A',\, \text{prof}_V\, A'')$, and if this is a strict inequality then $\text{prof}_V\, A' = \text{prof}_V\, A'' + 1$.

Proof. If all three of these modules have strictly positive profundities there is an element $f \in {}_V\mathfrak{m}$ which is not a zero-divisor for either A or A' or A'', hence for which $\{f\}$ is simultaneously an A-sequence, an A'-sequence, and an A''-sequence, as in the last paragraph of the proof of Theorem 18. The condition that $\{f\}$ is an A''-sequence can be restated as the condition that if $a \in A$ and $f\cdot a \in A'$ then $a \in A'$, or equivalently as the condition that $A' \cap f\cdot A = f\cdot A'$, where A' is viewed as a submodule of A; and in turn that implies that the induced sequence of ${}_V\mathcal{O}$-modules

$$0 \longrightarrow A'/f\cdot A' \longrightarrow A/f\cdot A \longrightarrow A''/f\cdot A'' \longrightarrow 0$$

is also exact. If the corollary holds for this latter exact sequence of ${}_V\mathcal{O}$-modules then it certainly holds for the original exact sequence of ${}_V\mathcal{O}$-modules, since $\text{prof}_V\,(A/f\cdot A) = \text{prof}_V\,A - 1$ and similarly for the other modules; and after repeating the argument as necessary it is clearly sufficient merely to prove the corollary in the special case that at least one of the three modules has zero profundity.

Suppose then that at least one of these three modules has zero profundity. If $\text{prof}_V\,A' = 0$ there is a nonzero element $a' \in A' \subseteq A$ such that ${}_V\mathfrak{m}\cdot a' = 0$, but then $\text{prof}_V\,A = 0$ as well. If $\text{prof}_V\,A = 0$ there is a nonzero element $a \in A$ such that ${}_V\mathfrak{m}\cdot a = 0$; if $a \in A'$ then $\text{prof}_V\,A' = 0$, while if $a \notin A'$ then the image of a in A'' is a nonzero element $a'' \in A''$ such that ${}_V\mathfrak{m}\cdot a'' = 0$ and hence $\text{prof}_V\,A'' = 0$. The only case still left to consider is that in which $\text{prof}_V\,A'' = 0$, $\text{prof}_V\,A' > 0$, and $\text{prof}_V\,A > 0$. In this final case there must exist a nonzero element $a'' \in A''$ such that ${}_V\mathfrak{m}\cdot a'' = 0$, or equivalently there must exist an element $a \in A$ such that $a \notin A'$ but ${}_V\mathfrak{m}\cdot a \subseteq A'$; and there must exist an element $f \in {}_V\mathfrak{m}$ which is not a zero-divisor for either A' or A. Then $f\cdot a \in A'$, $f\cdot a \notin f\cdot A'$, and ${}_V\mathfrak{m}f\cdot a \subseteq f\cdot A'$, so that $f\cdot a$ represents a nonzero element $\tilde{a} \in A'/f\cdot A'$ such that ${}_V\mathfrak{m}\cdot\tilde{a} = 0$; consequently $\text{prof}_V\,(A'/f\cdot A') = 0$, so that $\text{prof}_V\,A' = 1 = \text{prof}_V\,A'' + 1$. That suffices to complete the proof of the corollary.

At this point in the discussion it might be of interest to calculate the profundity of a useful specific example. Considering the regular local ring ${}_k\mathcal{O}$ as a module over itself, it is easy to see that $\mathrm{prof}_{{}_k\mathcal{O}}\, {}_k\mathcal{O} = k$, indeed that $\{z_1, \ldots, z_k\}$ is a maximal ${}_k\mathcal{O}$-sequence. For if $f_i \in {}_k\mathcal{O}$ are any germs of holomorphic functions such that $z_1f_1 + \ldots + z_\ell f_\ell = 0$ for some index ℓ then the product of each monomial in the Taylor expansion of the function f_ℓ by the variable z_ℓ must be divisible by at least one of the variables $z_1, z_2, \ldots, z_{\ell-1}$, from which it is apparent that $f_\ell \in {}_k\mathcal{O}\cdot z_1 + \ldots + {}_k\mathcal{O}\cdot z_{\ell-1}$; thus $\{z_1, \ldots, z_k\}$ is an ${}_k\mathcal{O}$-sequence. On the other hand $1 \in {}_k\mathcal{O}$ represents a nonzero element of ${}_k\mathcal{O}/{}_k\mathfrak{m}$, where of course ${}_k\mathfrak{m} = {}_k\mathcal{O}\cdot z_1 + \ldots + {}_k\mathcal{O}\cdot z_k$, and ${}_k\mathfrak{m}\cdot 1 = 0$ in ${}_k\mathcal{O}/{}_k\mathfrak{m}$; therefore $\mathrm{prof}_{{}_k\mathcal{O}}\, {}_k\mathcal{O}/{}_k\mathfrak{m} = 0$, so that $\{z_1, \ldots, z_k\}$ is a maximal ${}_k\mathcal{O}$-sequence.

(f) The concepts of homological dimension and profundity of ${}_V\mathcal{O}$-modules are closely related, and the analysis of this relationship sheds considerable light on both concepts. For any germ V of a complex analytic variety the local ring ${}_V\mathcal{O}$ can itself be viewed as an ${}_V\mathcal{O}$-module; the profundity of this module will be called simply the <u>profundity of the germ</u> V and will be denoted by $\mathrm{prof}\, V$, so that $\mathrm{prof}\, V = \mathrm{prof}_V\, {}_V\mathcal{O}$. With this notation the fundamental observation about the relationship between these two concepts is the following result of M Auslander and D. Buchsbaum.

Theorem 19. If A is a finitely generated ${}_V\mathcal{O}$-module for some germ V of a complex analytic variety and if $\text{hom dim}_V A < \infty$ then

$$\text{hom dim}_V A + \text{prof}_V A = \text{prof } V.$$

Proof. The proof is naturally by induction on $\text{hom dim}_V A$, but the first few cases are somewhat exceptional. First if $\text{hom dim}_V A = 0$ then $A \cong {}_V\mathcal{O}^r$ for some r, and the desired result in this case is that $\text{prof}_V\, {}_V\mathcal{O}^r = \text{prof}_V\, {}_V\mathcal{O} = \text{prof } V$. This is of course true when $r = 1$; and if it is true for some value r then applying Corollary 1 to Theorem 18 to the exact sequence of ${}_V\mathcal{O}$-modules

$$0 \longrightarrow {}_V\mathcal{O}^r \longrightarrow {}_V\mathcal{O}^{r+1} \longrightarrow {}_V\mathcal{O} \longrightarrow 0$$

it is evidently also true for the value $r+1$, and that suffices to prove the theorem in this case.

Next if $\text{hom dim}_V A = 1$ there is an exact sequence of ${}_V\mathcal{O}$-modules of the form

$$0 \longrightarrow {}_V\mathcal{O}^{r_1} \xrightarrow{\sigma_1} {}_V\mathcal{O}^r \xrightarrow{\sigma} A \longrightarrow 0\,.$$

In this case it suffices merely to show that $\text{prof}_V A < \text{prof } V$; for then applying Corollary 1 to Theorem 18 to this exact sequence it follows that $\text{prof}_V\, {}_V\mathcal{O}^{r_1} = \text{prof}_V A + 1$, hence that $\text{prof}_V A + 1 = \text{prof } V$ as desired. Suppose contrariwise that $\text{prof}_V A \geq \text{prof}_V\, {}_V\mathcal{O} = n$; then as in the last paragraph of the proof of Theorem 18 there are elements $f_i \in {}_V\mathcal{W}$ such that $\{f_1,\ldots,f_n\}$ is simultaneously an A-sequence and a maximal ${}_V\mathcal{O}$-sequence, and it follows readily that the induced sequence of ${}_V\mathcal{O}$-modules

$$0 \longrightarrow ({}_V\mathcal{O}/f_1\cdot{}_V\mathcal{O} + \ldots + f_n\cdot{}_V\mathcal{O})^{r_1} \xrightarrow{\tilde{\sigma}_1} ({}_V\mathcal{O}/f_1\cdot{}_V\mathcal{O} + \ldots + f_n\cdot{}_V\mathcal{O})^{r}$$

$$\xrightarrow{\tilde{\sigma}} A/f_1\cdot A + \ldots + f_n\cdot A \longrightarrow 0$$

is also exact. (The only nontrivial part is the injectivity of the homomorphism $\tilde{\sigma}_1$; but if $F \in {}_V\mathcal{O}^{r_1}$ and

$$\sigma_1(F) = f_1F_1 + \ldots + f_nF_n \in f_1\cdot{}_V\mathcal{O}^r + \ldots + f_n\cdot{}_V\mathcal{O}^r$$

then $0 = \sigma\sigma_1(F) = f_1\sigma(F_1) + \ldots + f_n\sigma(F_n)$, and since $\{f_1,\ldots,f_n\}$ is an A-sequence this in turn implies that $\sigma(F_i) = 0$ and hence that $F_i = \sigma_1(G_i)$ and
$F = f_1G_1 + \ldots f_nG_n \in f_1\cdot{}_V\mathcal{O}^{r_1} + \ldots + f_n\cdot{}_V\mathcal{O}^{r_1}$, since σ_1 is injective.) The homomorphism σ_1 can be represented by an $r\times r_1$ matrix $S = \{s_{ij}\}$ where $s_{ij} \in {}_V\mathcal{O}$, in the sense that $\sigma_1(F)$ is the matrix product SF when $F = \{f_j\} \in {}_V\mathcal{O}^{r_1}$ is viewed as a column vector of length r_1 formed of elements $f_j \in {}_V\mathcal{O}$; and the matrix S can be decomposed into the sum $S = S' + S''$ where $s'_{ij} \in \mathbb{C}$ and $s''_{ij} \in {}_V\mathcal{W}$. Now since
$\text{prof}_V\ ({}_V\mathcal{O}/f_1\cdot{}_V\mathcal{O} + \ldots + f_n\cdot{}_V\mathcal{O}) = 0$ there must exist an element $f \in {}_V\mathcal{O}$ such that $f \notin f_1\cdot{}_V\mathcal{O} + \ldots + f_n\cdot{}_V\mathcal{O}$ but ${}_V\mathcal{W}\cdot f \subseteq f_1\cdot{}_V\mathcal{O} + \ldots + f_n\cdot{}_V\mathcal{O}$. Then for any nonzero constant

column vector $C \in \mathbb{C}^{r_1}$ the product $fC \in {}_V\mathcal{O}^{r_1}$ represents a nonzero element of $({}_V\mathcal{O}/f_1 \cdot {}_V\mathcal{O} + \ldots + f_n \cdot {}_V\mathcal{O})^{r_1}$, and since $\tilde{\sigma}_1$ is injective $\sigma_1(fC) = fSC = fS'C + fS''C$ must consequently represent a nonzero element of $({}_V\mathcal{O}/f_1 \cdot {}_V\mathcal{O} + \ldots + f_n \cdot {}_V\mathcal{O})^r$; but $fS''C \in (f_1 \cdot {}_V\mathcal{O} + \ldots + f_n \cdot {}_V\mathcal{O})^r$ since $fs''_{ij} \in f \cdot {}_V\mathcal{W}$, and consequently $fS'C \neq 0$. Thus $S'C \neq 0$ for every nonzero vector $C \in \mathbb{C}^{r_1}$, and hence the constant matrix S' must be of rank r_1; but then after a suitable automorphism of ${}_V\mathcal{O}^r$ the matrix S can itself be reduced to the form $S = (S_1, 0)$ where S_1 is an invertible matrix of rank r_1 over the ring ${}_V\mathcal{O}$, and that means that $A \cong {}_V\mathcal{O}^{r-r_1}$ and hence that $\text{hom dim}_V A = 0$. That contradicts the assumption that $\text{hom dim}_V A = 1$, and hence suffices to conclude the proof of the theorem in this case.

Finally assume that the theorem has been proved for all finitely generated ${}_V\mathcal{O}$-modules of homological dimension strictly less than n for some integer $n \geq 2$; and consider a finitely generated ${}_V\mathcal{O}$-module A with $\text{hom dim}_V A = n$. There is then an exact sequence of ${}_V\mathcal{O}$-modules of the form

$$0 \longrightarrow \text{syz } A \longrightarrow {}_V\mathcal{O}^r \longrightarrow A \longrightarrow 0 ,$$

and $\text{hom dim}_V (\text{syz } A) = n - 1$ so the theorem holds for the module syz A. Thus

$$\text{prof}_V (\text{syz } A) = \text{prof } V - (n - 1) < \text{prof } V = \text{prof}_V {}_V\mathcal{O}$$

since $n \geq 2$, and hence it follows from Corollary 1 to Theorem 18 that $\text{prof}_V\,(\text{syz } A) = \text{prof}_V A + 1$; consequently $\text{prof}_V A = \text{prof } V - n$ as desired, and that suffices to conclude the proof of the whole theorem.

With results such as Theorem 19 in mind, the term homological codimension is sometimes used instead of profundity. The finiteness restriction in that theorem is essential since profundity is always finite but, as will shortly be seen, homological dimension is not necessarily finite; however there are cases in which the finiteness of the homological dimension can be guaranteed quite generally.

Theorem 20. Any finitely generated ${}_k\mathcal{O}$-module has finite homological dimension.

Proof. The proof is by induction on the dimension k; the case $k = 0$ is trivial since every module over ${}_0\mathcal{O} = \mathbb{C}$ is necessarily free, so assume that the theorem has been demonstrated for finitely generated ${}_{k-1}\mathcal{O}$-modules and let A be a finitely generated ${}_k\mathcal{O}$-module. The minimal free resolution of the module A can be split into two exact sequences of ${}_k\mathcal{O}$-modules

$$\cdots \longrightarrow {}_k\mathcal{O}^{r_2} \longrightarrow {}_k\mathcal{O}^{r_1} \longrightarrow A_1 \longrightarrow 0$$

$$0 \longrightarrow A_1 \longrightarrow {}_k\mathcal{O}^{r} \longrightarrow A \longrightarrow 0 ,$$

where the first of these is the minimal free resolution of $A_1 = \text{syz } A$. Since $A_1 \subseteq {}_k\mathcal{O}^{r}$ it follows that z_k is not a zero-divisor for either A_1 or ${}_k\mathcal{O}$, hence as noted several times before the induced sequence

$$\cdots \longrightarrow ({}_k\mathcal{O}/{}_k\mathcal{O}\cdot z_k)^{r_2} \longrightarrow ({}_k\mathcal{O}/{}_k\mathcal{O}\cdot z_k)^{r_1} \longrightarrow A_1/z_k\cdot A_1 \longrightarrow 0$$

must also be an exact sequence of ${}_k\mathcal{O}$-modules; actually of course since z_k annihilates all the modules in this sequence and ${}_k\mathcal{O}/{}_k\mathcal{O}\cdot z_k \cong {}_{k-1}\mathcal{O}$ the sequence can be viewed as the exact sequence of ${}_{k-1}\mathcal{O}$-modules

$$\cdots \longrightarrow {}_{k-1}\mathcal{O}^{r_2} \longrightarrow {}_{k-1}\mathcal{O}^{r_1} \longrightarrow A_1/z_k\cdot A_1 \longrightarrow 0 ,$$

hence as a free resolution of the ${}_{k-1}\mathcal{O}$-module $A_1/z_k\cdot A_1$. Note that in general if B is any finitely-generated ${}_k\mathcal{O}$-module and if $b_1,\ldots,b_n$ are elements of B such that the residue classes $\tilde{b}_i \in B/z_k\cdot B$ generate $B/z_k\cdot B$ as an ${}_{k-1}\mathcal{O}$-module then the elements b_i generate a submodule $B_1 \subseteq B$ such that $B = B_1 + z_k\cdot B = B_1 + {}_k\mathfrak{m}\cdot B$, and it follows from Nakayama's lemma that $B_1 = B$; therefore the minimal number of generators of $B/z_k\cdot B$ as an ${}_{k-1}\mathcal{O}$-module is the same as the minimal number of generators of B as an ${}_k\mathcal{O}$-module. In view of this observation the last exact sequence above must indeed be the minimal free resolution of the ${}_{k-1}\mathcal{O}$-module $A_1/z_k\cdot A_1$; but then it follows from the induction hypothesis that $r_n = 0$ whenever n is sufficiently large. Therefore the module A_1 and hence of course also the module A are of finite homological dimension, and the proof of the theorem is concluded.

The two preceding theorems can then be combined to refine the latter of them as follows. To simplify the notation $\text{hom dim}_k A$ will be used in place of $\text{hom dim}_{{}_k\mathcal{O}} A$ to denote the homological dimension of the ${}_k\mathcal{O}$-module A, and similarly $\text{prof}_k A$ will be used in place of $\text{prof}_{{}_k\mathcal{O}} A$ to denote the profundity of A.

<u>Corollary 1 to Theorem 20</u>. If A is a finitely generated ${}_k\mathcal{O}$-module then $0 \leq \text{hom dim}_k A \leq k$; if moreover $k > 0$ and there is an element of ${}_k\mathfrak{m}$ which is not a zero-divisor for the module A, as is the case when $A \subseteq {}_k\mathcal{O}^r$ for example, then $0 \leq \text{hom dim}_k A \leq k-1$.

Proof. Since $\text{hom dim}_k A < \infty$ as a consequence of Theorem 20 and since $\text{prof}\, \mathbb{C}^k = \text{prof}_{{}_k\mathcal{O}}\, {}_k\mathcal{O} = k$ as noted at the end of §3(e) it then follows from Theorem 19 that

$$\text{hom dim}_k A = \text{prof}\, \mathbb{C}^k - \text{prof}_k A = k - \text{prof}_k A \leq k ;$$

and if further there is an element of ${}_k\mathfrak{m}$ which is not a zero-divisor for A and $k > 0$ then $\text{prof}_k A \geq 1$, hence $\text{hom dim}_k A \leq k-1$. That serves to complete the proof of the corollary.

<u>Corollary 2 to Theorem 20</u>. For any germ V of a complex analytic variety $0 \leq \text{hom dim}\, V \leq \dim V - 1$, provided that $\dim V > 0$.

Proof. Since any finite analytic mapping $\varphi: V \to \mathbb{C}^k$ exhibits the local ring ${}_V\mathcal{O}$ as a finitely generated ${}_k\mathcal{O}$-module with no zero-divisors where $k = \dim V > 0$, it follows from Corollary 1 to Theorem 20 that $\text{hom dim}_\varphi V \leq k-1$; and consequently $\text{hom dim } V = \min_\varphi (\text{hom dim}_\varphi V) \leq k-1$ as desired, to complete the proof of the corollary.

Homological dimension and profundity of a germ V of a complex analytic variety refer to properties of the local ring ${}_V\mathcal{O}$, as an ${}_k\mathcal{O}$-module with respect to some finite analytic mapping $\varphi: V \to \mathbb{C}^k$ in the first instance and as an ${}_V\mathcal{O}$-module in the second instance; so in order to apply Theorem 19 to relate these two properties a further invariance property is required.

<u>Theorem 21</u>. If $\varphi: V_1 \to V_2$ is a finite analytic mapping between two germs of complex analytic varieties and A is a finitely generated ${}_{V_1}\mathcal{O}$-module then under the induced homomorphism $\varphi^*: {}_{V_2}\mathcal{O} \to {}_{V_1}\mathcal{O}$ the module A can also be viewed as a finitely generated ${}_{V_2}\mathcal{O}$-module and

$$\text{prof}_{V_1} A = \text{prof}_{V_2} A .$$

Proof. If $\{f_1,\ldots,f_n\}$ with $f_i \in {}_{V_2}\mathcal{W}$ is a maximal A-sequence when A is viewed as an ${}_{V_2}\mathcal{O}$-module then since the action of an element $f \in {}_{V_2}\mathcal{O}$ on A is defined as the action of the element $\varphi^*(f) \in {}_{V_1}\mathcal{O}$ on the ${}_{V_1}\mathcal{O}$-module A it is apparent that $\{\varphi^*(f_1),\ldots,\varphi^*(f_n)\}$ is also an A-sequence when A is viewed as an ${}_{V_1}\mathcal{O}$-module; hence $\text{prof}_{V_1} A \geq \text{prof}_{V_2} A = n$. On the other hand the ${}_{V_2}\mathcal{O}$-module

$$\tilde{A} = A/(f_1\cdot A + \ldots + f_n\cdot A) = A/(\varphi^*(f_1)\cdot A + \ldots + \varphi^*(f_n)\cdot A)$$

has the property that $\text{prof}_{V_2}\ \tilde{A} = 0$, hence there is a nonzero element $\tilde{a} \in \tilde{A}$ such that $\varphi^*({}_{V_2}\mathcal{W})\cdot\tilde{a} = 0$. Now any element $f \in {}_{V_1}\mathcal{W}$ is necessarily integral over the submodule $\varphi^*({}_{V_2}\mathcal{O}) \subseteq {}_{V_1}\mathcal{O}$, so there are elements $f_i \in {}_{V_2}\mathcal{O}$ such that

$$f^r + \varphi^*(f_1)f^{r-1} + \ldots + \varphi^*(f_r) = 0 ;$$

and it can even be assumed that $f_i \in {}_{V_2}\mathcal{W}$ for $1 \leq i \leq r$. (If the germ V_2 is represented by a germ V_2 of a complex analytic subvariety at the origin in $\mathbb{C}^k$ and $\varphi^*(f_i) = g_i = G_i|V_2$ for some germs $G_i \in {}_k\mathcal{O}$ then by the Weierstrass preparation and division theorems the polynomial

$$P(X) = X^r + G_1X^{r-1} + \ldots + G_r \in {}_k\mathcal{O}[X]$$

can be written as the product $P(X) = P_1(X)\cdot P_2(X)$ of a Weierstrass polynomial $P_1(X) \in {}_k\mathcal{O}[X]$ and a polynomial $P_2(X) \in {}_k\mathcal{O}[X] \subseteq {}_{k+1}\mathcal{O}$ such that $P_2(X)$ is a unit in ${}_{k+1}\mathcal{O}$ or equivalently such that the constant term in the polynomial $P_2(X)$ is a unit in ${}_k\mathcal{O}$. Since $0 = P(F)|V_2 = (P_1(F)|V_2)\cdot(P_2(F)|V_2)$ where $F|V_2 = f$ it follows that either $P_1(F)|V_2 = 0$ or $P_2(F)|V_2 = 0$; but since the constant term in $P_2(F)$ does not vanish at the base point of the germ V_2 while the function $F|V_2 = f \in {}_{V_2}\mathcal{W}$ does vanish there, it is impossible that $P_2(F)|V_2 = 0$ hence necessarily $P_1(F)|V_2 = 0$. The polynomial $P(X)$ can then be replaced by $P_1(X)$, hence it can be assumed that $f_i \in {}_{V_2}\mathcal{M}$ for $1 \leq i \leq r$ as desired.) Since $\varphi^*(f_i)\cdot\tilde{a} \in \varphi^*({}_{V_2}\mathcal{M})\cdot\tilde{a} = 0$ it then follows that $f^r\cdot\tilde{a} = 0$; hence there is some integer s with $1 \leq s \leq r$ for which $f\cdot f^{s-1}\tilde{a} = 0$ but $f^{s-1}\tilde{a} \neq 0$, so that f must be a zero-divisor for the ${}_{V_1}\mathcal{O}$-module A. Since this is true for every $f \in {}_{V_1}\mathcal{M}$ it follows that $\operatorname{prof}_{V_1} \tilde{A} = 0$, and consequently $\operatorname{prof}_{V_1} A \leq n = \operatorname{prof}_{V_2} A$ as well. That suffices to conclude the proof of the theorem.

The combination of this and the preceding two theorems yields a number of useful and interesting consequences, with which these notes will conclude.

<u>Corollary 1 to Theorem 21</u>. If V is a germ of a complex analytic variety with $\dim V = k$ then

$$\text{hom dim } V + \text{prof } V = k .$$

Moreover if $\varphi: V \to \mathbb{C}^k$ is any finite analytic mapping then

$$\text{hom dim}_\varphi V = \text{hom dim } V.$$

Proof. For any finite analytic mapping $\varphi: V \to \mathbb{C}^k$ it follows from Theorem 21 that $\text{prof } V = \operatorname{prof}_V {}_V\mathcal{O} = \operatorname{prof}_\varphi {}_V\mathcal{O}$, where $\operatorname{prof}_\varphi {}_V\mathcal{O}$ denotes the profundity of ${}_V\mathcal{O}$ when considered as exhibited as an ${}_k\mathcal{O}$-module by the analytic mapping φ; and since $\text{prof } \mathbb{C}^k = k$ as observed at the end of §3(e) while ${}_V\mathcal{O}$ is of finite homological dimension as an ${}_k\mathcal{O}$-module as a consequence of Theorem 20, it then follows from Theorem 19 that

$\text{prof}_{\varphi}\, {}_V\mathcal{O} = k - \text{hom dim}_{\varphi}\, {}_V\mathcal{O} = k - \text{hom dim}_{\varphi} V$, and consequently $\text{hom dim}_{\varphi} V + \text{prof } V = k$. On the one hand there is a finite analytic mapping φ for which $\text{hom dim}_{\varphi} V = \text{hom dim } V$, and hence $\text{hom dim } V + \text{prof } V = k$; but on the other hand the expression $\text{hom dim}_{\varphi} V = k - \text{prof } V$ is independent of the choice of the mapping φ, so that $\text{hom dim}_{\varphi} V = \text{hom div } V$ for any φ. That suffices to conclude the proof of the corollary.

It of course follows from this that if V is a germ of complex analytic variety of dimension k then all finite analytic mappings $\varphi: V \to \mathbb{C}^k$ exhibit ${}_V\mathcal{O}$ as finitely generated ${}_k\mathcal{O}$-modules having the same homological dimension, this common value being called the homological dimension of the germ V; this simplifies the definition given at the beginning of §3(d). Note further that if $\dim V = k$ and $\varphi: V \to \mathbb{C}^n$ is a finite analytic mapping for some $n \geqq k$ then φ can be written as the composition of a finite analytic mapping $\varphi_1: V \to \mathbb{C}^k$ and a finite analytic mapping $\varphi_2: \mathbb{C}^k \to \mathbb{C}^n$; then from Theorem 21 it follows that $\text{prof } V = \text{prof}_k V = \text{prof}_n V$, and hence by Theorem 19 it is also true that $k - \text{hom dim}_{\varphi_1} V = n - \text{hom dim}_{\varphi} V$. Thus $\text{hom dim}_{\varphi} V = \text{hom dim } V + (n - k)$.

<u>Corollary 2 to Theorem 21</u>. If V is a perfect germ of a complex analytic variety with $\dim V = k$ then every finite analytic mapping $\varphi: V \to \mathbb{C}^k$ exhibits ${}_V\mathcal{O}$ as a free ${}_k\mathcal{O}$-module.

Proof. This is merely a special case of Corollary 1 to Theorem 21.

Corollary 3 to Theorem 21. For any germ V of a complex analytic variety and any integer $d > 0$ the subset $\{p \in V \mid \text{hom dim } V_p \geqq d\}$ is a proper complex analytic subvariety of V.

Proof. If $\varphi: V \to \mathbb{C}^k$ is any finite analytic mapping where $k = \dim V$ then from Corollary 1 to Theorem 21 it follows that $\text{hom dim } V_p = \text{hom dim}_{{}_k\mathcal{O}_z} {}_V\mathcal{O}_p$ for every point $p \in V$ sufficiently near the base point of V, where ${}_V\mathcal{O}_p$ is exhibited as an ${}_k\mathcal{O}_z$-module with $z = \varphi(p)$ by the mapping φ. Now the direct image sheaf $\varphi_*({}_V\mathcal{O})$ is a coherent analytic sheaf in an open neighborhood U of the origin in $\mathbb{C}^k$ and $\varphi_*({}_V\mathcal{O})_z = {}_V\mathcal{O}_{p_1} \oplus \ldots \oplus {}_V\mathcal{O}_{p_n}$ for any $z \in U$, where $\varphi^{-1}(z) = \{p_1, \ldots, p_n\}$ and ${}_V\mathcal{O}_{p_i}$ are exhibited as ${}_k\mathcal{O}_z$-modules by the mapping φ. Since $\text{syz}^j (A \oplus B) \cong \text{syz}^j A \oplus \text{syz}^j B$ for any finitely generated ${}_k\mathcal{O}_z$-modules A, B by Corollary 1 to Theorem 16, it is easy to see that $\text{hom dim}_{{}_k\mathcal{O}_z} \varphi_*({}_V\mathcal{O})_z \leq d - 1$ precisely when $\text{hom dim}_{{}_k\mathcal{O}_z} {}_V\mathcal{O}_{p_i} \leq d - 1$ for all i with $1 \leq i \leq n$, or equivalently that $\text{hom dim}_{{}_k\mathcal{O}_z} \varphi_*({}_V\mathcal{O})_z \geq d$ precisely when $\text{hom dim}_{{}_k\mathcal{O}_z} {}_V\mathcal{O}_{p_i} \geq d$ for some i with $1 \leq i \leq n$; therefore the image of the subset $S_d = \{p \in V \mid \text{hom dim } V_p \geq d\} \subseteq V$ under the mapping $\varphi: V \to \mathbb{C}^k$ is precisely the set $\varphi(S_d) = \{z \in U \mid \text{hom dim}_{{}_k\mathcal{O}_z} \varphi_*({}_V\mathcal{O})_z \geq d\}$, and if $d > 0$ this is a proper analytic subvariety of U as a consequence of Corollary 3 to Theorem 16. Consequently the image of the subset $S_d \subseteq V$ under any finite analytic mapping $\varphi: V \to \mathbb{C}^k$ where $k = \dim V$ is a proper complex analytic subvariety of an open neighborhood of the origin in $\mathbb{C}^k$ if $d > 0$. It is easy to see from this that S_d itself must then be a proper analytic subvariety of V. The intersection of the subvarieties $\varphi^{-1}\varphi(S_d) \subset V$ for all finite analytic mappings $\varphi\colon V \to \mathbb{C}^k$ is the germ of a proper analytic subvariety $W \subset V$ such that $S_d \subseteq W$; and for any point $p \in V - W$ sufficiently near the base point of V choose a finite analytic mapping $\varphi\colon V \to \mathbb{C}^k$ such that all points of $\varphi^{-1}\varphi(p)$ except p are regular points of V, noting then that

$d > \text{hom dim}_{{}_k\mathcal{O}_z} \varphi_*({}_V\mathcal{O})_z = \text{hom dim}_{{}_k\mathcal{O}_z} {}_V\mathcal{O}_p$ where $z = \varphi(p)$

hence that $p \notin S_d$ and therefore that $W \subseteq S_d$. That suffices to conclude the proof of the corollary.

Corollary 4 to Theorem 21. If V is a germ of a complex analytic variety with $\dim V = k$ and $\text{hom dim}\, V = d$, and if $\{f_1,\ldots,f_n\}$ is an ${}_V\mathcal{O}$-sequence for some elements $f_i \in {}_V\mathcal{W}$ and $\mathfrak{A} = {}_V\mathcal{O}\cdot f_1 + \ldots + {}_V\mathcal{O}\cdot f_n \subseteq {}_V\mathcal{O}$ is the ideal generated by these elements, then $W = \text{loc}\,\mathfrak{A}$ is a complex analytic subvariety of V with $\dim W = k - n$; if moreover $\mathfrak{A}$ is a radical ideal then $\text{hom dim}\, W = d$.

Proof. The first assertion is easily demonstrated by induction on the index n. For the case $n = 1$ the condition that $\{f_1\}$ be an ${}_V\mathcal{O}$-sequence is just the condition that f_1 not be a zero-divisor in the ring ${}_V\mathcal{O}$; and it then follows from Theorem 9(e) of CAV I that $\dim(\text{loc}\,{}_V\mathcal{O}\cdot f_1) = k - 1$ as desired. Assuming that the result has been demonstrated for the case $n - 1$ and considering the ${}_V\mathcal{O}$-sequence $\{f_1,\ldots,f_n\}$, the ideal

$\mathfrak{A}' = {}_V\mathcal{O}\cdot f_1 + \ldots + {}_V\mathcal{O}\cdot f_{n-1} \subseteq {}_V\mathcal{O}$ has the property that $W' = \operatorname{loc} \mathfrak{A}'$ is a complex analytic subvariety of V with $\dim W' = k - n + 1$; and in view of the case $n = 1$ already established, in order to complete the proof of the desired result it is only necessary to show that the restriction $f_n|W'$ is not a zero-divisor in the ring ${}_{W'}\mathcal{O}$, or equivalently that f_n is not a zero-divisor for the ${}_V\mathcal{O}$-module ${}_V\mathcal{O}/\sqrt{\mathfrak{A}'}$. If there were an element $g \in {}_V\mathcal{O}$ such that $g \notin \sqrt{\mathfrak{A}'}$ but $gf_n \in \sqrt{\mathfrak{A}'}$ then clearly there would also be an element $h \in {}_V\mathcal{O}$ such that $h \notin \mathfrak{A}'$ but $hf_n \in \mathfrak{A}'$, since $g^\nu f_n^\nu \in \mathfrak{A}'$, $g^\nu \notin \mathfrak{A}'$ for some integer $\nu \geq 1$; but then f_n would be a zero-divisor for the ${}_V\mathcal{O}$-module ${}_V\mathcal{O}/\mathfrak{A}'$, in contradiction to the assumption that $\{f_1,\ldots,f_n\}$ is an ${}_V\mathcal{O}$-sequence. Turning then to the second assertion, if $\mathfrak{A}$ is a radical ideal, $\mathfrak{A} = \operatorname{id} W$ and ${}_V\mathcal{O}/\mathfrak{A} \cong {}_W\mathcal{O}$; and the structure of ${}_W\mathcal{O}$ as an ${}_V\mathcal{O}$-module is just that induced by the inclusion mapping $W \to V$, so since this is a finite analytic mapping it follows from Theorem 21 that $\operatorname{prof} W = \operatorname{prof}_W {}_W\mathcal{O} = \operatorname{prof}_V {}_W\mathcal{O} = \operatorname{prof}_V {}_V\mathcal{O}/\mathfrak{A}$. Since $\{f_1,\ldots,f_n\}$ is an ${}_V\mathcal{O}$-sequence it is also apparent that $\operatorname{prof}_V {}_V\mathcal{O}/\mathfrak{A} = \operatorname{prof}_V {}_V\mathcal{O} - n = \operatorname{prof} V - n$, hence $\operatorname{prof} W = \operatorname{prof} V - n$. Then applying Corollary 1 to Theorem 21 it follows that $\operatorname{prof} V = k - d$ and $\operatorname{hom\,dim} W = \dim W - \operatorname{prof} W = (k - n) - (k - d - n) = d$; and that suffices to conclude the proof of the corollary.

It is convenient to say that a subvariety W of a germ V of a complex analytic variety is a <u>complete intersection in</u> V if the ideal id $W \subseteq {}_V\mathcal{O}$ is generated by elements $f_1,\ldots,f_n$ of ${}_V\mathcal{W}$ such that $\{f_1,\ldots,f_n\}$ is an ${}_V\mathcal{O}$-sequence; in such a case it follows from Corollary 4 to Theorem 21 that dim W = dim V - n and hom dim W = hom dim V. Since hom dim W $\leq$ dim W - 1 for any complex analytic variety W with dim W > 0 as a consequence of Corollary 2 to Theorem 20, it is apparent that a subvariety $W \subseteq V$ for which $0 <$ dim W $\leq$ hom dim V can never be a complete intersection in V. In particular in the extreme case that hom dim V = dim V - 1 no proper positive-dimensional complex analytic subvariety of V can be a complete intersection in V; thus if hom dim V = dim V - 1 and if $f \in {}_V\mathcal{W}$ is not a zero-divisor in the ring ${}_V\mathcal{O}$ then ${}_V\mathcal{O}\cdot f \subset \sqrt{{}_V\mathcal{O}\cdot f}$. In the other extreme case of a perfect germ V of a complex analytic variety this dimensional restriction disappears; and every subvariety of V which is a complete intersection in V is also a perfect germ of a complex analytic variety. For a pure-dimensional germ V of a complex analytic variety this definition can be simplified somewhat, since it is easy to see that $\{f_1,\ldots,f_n\}$ is an ${}_V\mathcal{O}$-sequence whenever $f_i \in {}_V\mathcal{W}$ are elements generating an ideal $\mathcal{U} \subseteq {}_V\mathcal{O}$ such that dim loc $\mathcal{U}$ = dim V - n. (It is apparent from Theorem 9(f) of CAV I that if dim loc $\mathcal{U}$ = dim V - n and if $\mathcal{U}_i$ denotes the ideal in ${}_V\mathcal{O}$ generated by the elements $f_1,\ldots,f_i$ for $1 \leq i \leq n$ then loc $\mathcal{U}_i$ is a pure-dimensional subvariety of V and dim loc $\mathcal{U}_i$ = dim V - i. If f_{i+1} were a zero-divisor for the module ${}_V\mathcal{O}/\mathcal{U}_i$ then f_{i+1} would have to vanish identically on some irreducible component of loc $\mathcal{U}_i$; and that would imply that $\dim \operatorname{loc} \mathfrak{A}_{i+1} = \dim \operatorname{loc} \mathfrak{A}_i = \dim V - i$, which is impossible.) Thus a subvariety W of a pure-dimensional germ V of a complex analytic variety is a complete intersection in V if and only if the ideal $\operatorname{id} W \subseteq {}_V\mathcal{O}$ is generated by n elements where $n = \dim V - \dim W$. It is traditional merely to say that a germ V of a complex analytic variety is a complete intersection if it can be represented as a complete intersection in a regular germ of a complex analytic variety. Any complete intersection is consequently a perfect germ of a complex analytic variety; the converse is of course not true, since arbitrary one-dimensional germs of complex analytic varieties are perfect as a consequence of Corollary 2 to Theorem 20 but are not necessarily complete intersections.

Corollary 5 to Theorem 21. If $\varphi: V_1 \to V_2$ is a finite analytic mapping between two germs of complex analytic varieties and if φ exhibits ${}_{V_1}\mathcal{O}$ as a finitely generated ${}_{V_2}\mathcal{O}$-module such that $\operatorname{hom\,dim}_{V_2} {}_{V_1}\mathcal{O} < \infty$ then

$$\operatorname{hom\,dim} V_1 - \dim V_1 = \operatorname{hom\,dim} V_2 - \dim V_2 + \operatorname{hom\,dim}_{V_2} {}_{V_1}\mathcal{O} .$$

Proof. It follows from Corollary 1 to Theorem 21 that $\operatorname{hom\,dim} V_i = \dim V_i - \operatorname{prof} V_i$; and it follows from Theorem 21 itself that $\operatorname{prof} V_1 = \operatorname{prof}_{V_1} {}_{V_1}\mathcal{O} = \operatorname{prof}_{V_2} {}_{V_1}\mathcal{O}$, while from Theorem 19 then $\operatorname{prof}_{V_2} {}_{V_1}\mathcal{O} = \operatorname{prof} V_2 - \operatorname{hom\,dim}_{V_2} {}_{V_1}\mathcal{O}$. Combining these observations,

$$\begin{aligned}\text{hom dim } V_1 &= \dim V_1 - \text{prof } V_1 \\ &= \dim V_1 - (\text{prof } V_2 - \text{hom dim}_{V_2}\, {}_{V_1}\mathcal{O}) \\ &= \dim V_1 - \dim V_2 + \text{hom dim } V_2 + \text{hom dim}_{V_2}\, {}_{V_1}\mathcal{O}\end{aligned}$$

as desired, and the proof of the corollary is thereby concluded.

In particular if V_1, V_2 are pure-dimensional germs of complex analytic varieties of the same dimension and if $\varphi: V_1 \to V_2$ is a simple analytic mapping exhibiting ${}_{V_1}\mathcal{O}$ as a finitely generated ${}_{V_2}\mathcal{O}$-module such that $\text{hom dim}_{V_2}\, {}_{V_1}\mathcal{O} < \infty$ then $\text{hom dim } V_1 = \text{hom dim } V_2 + \text{hom dim}_{V_2} V_1$. Note that $\text{hom dim}_{V_2} V_1 = 0$ only when ${}_{V_1}\mathcal{O}$ is a free ${}_{V_2}\mathcal{O}$-module, indeed a free ${}_{V_2}\mathcal{O}$-module of rank 1 since φ is simple, hence only when ${}_{V_1}\mathcal{O} \cong {}_{V_2}\mathcal{O}$; thus if V_1 and V_2 are not equivalent germs of complex analytic varieties then $\text{hom dim } V_1 > \text{hom dim } V_2$. Therefore if V_1 is a perfect germ of a complex analytic variety and if $\varphi: V_1 \to V_2$ is a simple analytic mapping which is not an equivalence then $\text{hom dim}_{V_2}\, {}_{V_1}\mathcal{O} = \infty$; this provides a very natural class of examples of finitely generated ${}_{V_2}\mathcal{O}$-modules which do not have finite homological dimension.

Appendix. Local cohomology groups of complements of complex analytic subvarieties.

The investigation of the local cohomology groups of complements of complex analytic subvarieties is an interesting and important topic in the study of complex analytic varieties, and merits a detailed separate treatment; however the discussion of a few simple results in that direction will be appended here, to complete the considerations in §3(d) for those readers not familiar with that topic. No attempt will be made here to review the general properties of cohomology groups with coefficients in a coherent analytic sheaf; for that the reader can be referred to such texts as L. Hörmander, An Introduction to Complex Analysis in Several Variables, or R. C. Gunning and H. Rossi, Analytic Functions of Several Complex Variables. In section 4.3 of the first reference or section VI.D of the second reference the cohomology groups $H^p(D, \mathcal{S})$ of a paracompact Hausdorff space D with coefficients in a sheaf $\mathcal{S}$ of abelian groups are expressed in terms of the cohomology groups $H^p(\mathcal{U}, \mathcal{S})$ of coverings $\mathcal{U} = \{U_i\}$ of the space D; indeed Leray's theorem on cohomology (Theorem VI,D4 of the second reference) describes conditions under which there are isomorphisms $H^p(D, \mathcal{S}) \cong H^p(\mathcal{U}, \mathcal{S})$. It is convenient to have at hand a slight extension of that theorem, as in the following lemma; the proof follows almost precisely the proof of Leray's theorem in the second reference noted above, hence will be omitted altogether here.

Lemma 1. If $\mathcal{S}$ is a sheaf of abelian groups on a paracompact Hausdorff space D and if $\mathfrak{U} = \{U_i\}$ is a covering of D by open sets U_i such that

$$H^p(U_{i_0} \cap \ldots \cap U_{i_m}, \mathcal{S}) = 0 \quad \text{whenever} \quad 1 \leq p \leq r$$

for any finite intersection of the sets in $\mathfrak{U}$, then

$$H^p(D, \mathcal{S}) \cong H^p(\mathfrak{U}, \mathcal{S}) \quad \text{whenever} \quad 0 \leq p \leq r .$$

The more detailed results which will be treated here are primarily simple consequences of the following lemma, which is itself a special case of a result of J. Frenkel (Bull. Soc. Math. France, vol. 85, 1957, pp. 135-230).

Lemma 2. For the open subset $U \subseteq \mathbb{C}^n$ defined by

$$U = \left\{ (z_1, \ldots, z_n) \in \mathbb{C}^n \,\middle|\, \begin{array}{l} |z_1| < \delta_1, \ \ldots, \ |z_d| < \delta_d, \\ (z_1, \ldots, z_d) \neq (0, \ldots, 0), \\ (z_{d+1}, \ldots, z_n) \in D \end{array} \right\}$$

where $3 \leq d \leq n$, $0 < \delta_1, \ldots, \delta_d < \infty$, and $D \subseteq \mathbb{C}^{n-d}$ is a domain of holomorphy, it follows that

$$H^p(U, \mathcal{O}) = 0 \qquad \text{whenever} \quad 1 \leq p \leq d-2.$$

Proof. The open subsets

$$U_i = \left\{ (z_1, \ldots, z_n) \in \mathbb{C}^n \,\middle|\, \begin{array}{l} |z_1| < \delta_1, \ \ldots, \ |z_d| < \delta_d, \ z_i \neq 0 \\ (z_{d+1}, \ldots, z_n) \in D \end{array} \right\}$$

for $1 \leq i \leq d$ clearly form a covering $\mathfrak{U} = \{U_1,\ldots,U_d\}$ of the set U; and since the sets U_i and all their intersections are domains of holomorphy and hence have trivial analytic cohomology groups in all positive dimensions it follows from Lemma 1 that $H^p(U,\mathcal{O}) \cong H^p(\mathfrak{U},\mathcal{O})$ for all p. The cohomology groups $H^p(\mathfrak{U},\mathcal{O})$ will here be considered as being defined by skew-symmetric cochains. If $i,i_0,\ldots,i_p$ are any distinct indices with $1 \leq i,i_0,\ldots,i_p \leq d$ and f is any holomorphic function on the set $U_i \cap U_{i_0} \cap \ldots \cap U_{i_p}$ then the function f can be expanded in a Laurent series of the form

$$f(z_1,\ldots,z_n) = \sum_{\nu=-\infty}^{+\infty} f_\nu(z_1,\ldots,z_{i-1},z_{i+1},\ldots,z_n)z_i^\nu$$

where the coefficients $f_\nu(z_1,\ldots,z_{i-1},z_{i+1},\ldots,z_n)$ are holomorphic in the projection of the set U_i to the space $\mathbb{C}^{n-1}$. Setting

$$R_if(z_1,\ldots,z_n) = \sum_{\nu=0}^{+\infty} f_\nu(z_1,\ldots,z_{i-1},z_{i+1},\ldots,z_n)z_i^\nu$$

then defines a linear mapping

$$R_i\colon \Gamma(U_i \cap U_{i_0} \cap \ldots \cap U_{i_p},\mathcal{O}) \longrightarrow \Gamma(U_{i_0} \cap \ldots \cap U_{i_p},\mathcal{O})\ ;$$

and this can be used in turn to define a linear mapping

$$Q_i\colon C^p(\mathfrak{U},\mathcal{O}) \longrightarrow C^{p-1}(\mathfrak{U},\mathcal{O})$$

by setting

$$(Q_if)(U_{i_0},\ldots,U_{i_{p-1}}) = (R_if)(U_i,U_{i_0},\ldots,U_{i_{p-1}})$$

for any skew-symmetric cochain $f \in C^p(\mathfrak{U},\mathcal{O})$, noting that $(Q_i f)(U_{i_0},\ldots,U_{i_{p-1}}) = 0$ unless the indices $i, i_0, \ldots, i_{p-1}$ are distinct. Finally define the linear mapping

$$P_i : C^p(\mathfrak{U},\mathcal{O}) \longrightarrow C^p(\mathfrak{U},\mathcal{O})$$

by setting

$$P_i f = f - \delta Q_i f - Q_i \delta f$$

for any skew-symmetric cochain $f \in C^p(\mathfrak{U},\mathcal{O})$. Now if f is a cocycle it is clear that $P_i f$ is also a cocycle, indeed that the cocycles $P_i f$ and f are cohomologous. On the other hand though

$$\begin{aligned}
&(P_i f)(U_{i_0},\ldots,U_{i_p}) \\
&= f(U_{i_0},\ldots,U_{i_p}) - (\delta Q_i f)(U_{i_0},\ldots,U_{i_p}) - (Q_i \delta f)(U_{i_0},\ldots,U_{i_p}) \\
&= f(U_{i_0},\ldots,U_{i_p}) - \sum_{k=0}^{p} (-1)^k (Q_i f)(U_{i_0},\ldots,\hat{U}_{i_k},\ldots,U_{i_p}) \\
&\qquad - R_i(\delta f)(U_i,U_{i_0},\ldots,U_{i_p}) \\
&= f(U_{i_0},\ldots,U_{i_p}) - \sum_{k=0}^{p} (-1)^k R_i f(U_i,U_{i_0},\ldots,\hat{U}_{i_k},\ldots,U_{i_p}) \\
&\qquad - R_i[f(U_{i_0},\ldots,U_{i_p}) - \sum_{k=0}^{p} (-1)^k f(U_i,U_{i_0},\ldots,\hat{U}_{i_k},\ldots,U_{i_p})] \\
&= f(U_{i_0},\ldots,U_{i_p}) - R_i f(U_{i_0},\ldots,U_{i_p}) ;
\end{aligned}$$

and since it is clear that $R_i f(U_{i_0},\ldots,U_{i_p}) = f(U_{i_0},\ldots,U_{i_p})$ when the indices $i, i_0, \ldots, i_p$ are distinct, it follows that $(P_i f)(U_{i_0}, \ldots, U_{i_p}) = 0$ whenever the indices $i, i_0, \ldots, i_p$ are distinct or equivalently that $(P_i f)(U_{i_0}, \ldots, U_{i_p}) \neq 0$ only when $i \in \{i_0, \ldots, i_p\}$ Then upon repeating this observation it is apparent that for any skew-symmetric cocycle $f \in C^p(\mathcal{U}, \mathcal{O})$ the cocycles $Pf = P_1 P_2 \cdots P_d f \in C^p(\mathcal{U}, \mathcal{O})$ and f are cohomologous, and that $(Pf)(U_{i_0}, \ldots, U_{i_p}) \neq 0$ only when $\{1, \ldots, d\} \subseteq \{i_0, \ldots, i_p\}$; the cocycle Pf is consequently trivial unless $p + 1 \geqq d$, hence any skew-symmetric cocycle $f \in C^p(\mathcal{U}, \mathcal{O})$ is cohomologous to zero if $1 \leq p \leq d - 2$, and the proof of the lemma is thereby concluded.

The principal consequence of this result which is of interest here is the following.

<u>Theorem 2 2</u>. If D is an open subset of $\mathbb{C}^n$ such that $H^p(D, \mathcal{O}) = 0$ whenever $1 \leq p \leq d - 2$ for some integer $d \geqq 3$, and if V is a complex analytic subvariety of D such that $\dim V \leq n - d$, then

$$H^p(D - V, \mathcal{O}) = 0 \quad \text{whenever} \quad 1 \leq p \leq d - 2 .$$

Proof. Note that in order to prove the theorem it is sufficient merely to show that there is a covering $\mathcal{U} = \{U_i\}$ of the set D by open domains of holomorphy U_i such that

$$(4) \quad H^p(U_{i_0} \cap \ldots \cap U_{i_m} \cap (D - V), \mathcal{O}) = 0 \quad \text{whenever} \quad 1 \leq p \leq d - 2$$

for any finite intersection $U_{i_0} \cap \ldots \cap U_{i_m}$ of the sets in $\mathcal{U}$.

Indeed since the intersections $U_{i_0} \cap \dots \cap U_{i_m}$ are also domains of holomorphy and hence have trivial analytic cohomology groups in positive dimensions, it follows from Lemma 1 that $H^p(\mathfrak{U},\mathcal{O}) \cong H^p(D,\mathcal{O})$, and consequently $H^p(\mathfrak{U},\mathcal{O}) = 0$ whenever $1 \leq p \leq d-2$; the sets $V_i = U_i \cap (D-V) = U_i - U_i \cap V$ form an open covering $\mathfrak{V} = \{V_i\}$ of the set $D-V$, and since (4) is precisely the condition that Lemma 1 apply to the covering $\mathfrak{V}$ it follows that $H^p(\mathfrak{V},\mathcal{O}) \cong H^p(D-V, \mathcal{O})$ whenever $1 \leq p \leq d-2$. Any cocycle $f \in C^p(\mathfrak{V},\mathcal{O})$ consists of sections $f(V_{i_0},\dots,V_{i_p}) \in \Gamma(V_{i_0} \cap \dots \cap V_{i_p}, \mathcal{O})$, and since $V_{i_0} \cap \dots \cap V_{i_p}$ is the complement of an analytic subvariety of codimension at least 3 in the set $U_{i_0} \cap \dots \cap U_{i_p}$ it follows from the extended Riemann removable singularities theorem that the function $f(V_{i_0},\dots,V_{i_p})$ extends to a holomorphic function $\tilde{f}(U_{i_0},\dots,U_{i_p}) \in \Gamma(U_{i_0} \cap \dots \cap U_{i_p}, \mathcal{O})$ and hence that the cocycle $f \in C^p(\mathfrak{V},\mathcal{O})$ extends to a cocycle $\tilde{f} \in C^p(\mathfrak{U},\mathcal{O})$; but if $1 \leq p \leq d-2$ there exists a cochain $\tilde{g} \in C^p(\mathfrak{U},\mathcal{O})$ such that $\tilde{f} = \delta\tilde{g}$, and the restriction of g determines a cochain $g \in C^p(\mathfrak{V},\mathcal{O})$ such that $f = \delta g$. Therefore $H^p(D-V,\mathcal{O}) \cong H^p(\mathfrak{V},\mathcal{O}) = 0$ whenever $1 \leq p \leq d-2$ as desired.

To apply these observations, consider first the special case that D is a domain of holomorphy in $\mathbb{C}^n$ and V is the linear subvariety

$$V = \{(z_1,\dots,z_n) \in D \mid z_1 = \dots = z_d = 0\} .$$

For any point $a \in V$ choose a polydisc $U_a \subseteq D$ centered at the point a; note that any finite intersection of such polydiscs is a set of the form

$$U_{a_0} \cap \ldots \cap U_{a_m}$$

$$= \{(z_1,\ldots,z_n) \in \mathbb{C}^n \mid \ |z_1| < \delta_1, \ \ldots, \ |z_d| < \delta_d, \ (z_{d+1},\ldots,z_n) \in D'\}$$

where D' is a domain of holomorphy in $\mathbb{C}^{n-d}$, and hence as a consequence of Lemma 2 that

$$H^p(U_{a_0} \cap \ldots \cap U_{a_m} \cap (D - V), \mathcal{O}) = 0 \quad \text{whenever} \quad 1 \leq p \leq d-2 .$$

These polydiscs U_a, together with a number of polydiscs contained in D and not intersecting V at all, form a covering $\mathcal{U} = \{U_i\}$ of the set D by domains of holomorphy; and if $U_{i_0} \cap \ldots \cap U_{i_m}$ is a finite intersection of sets in $\mathcal{U}$ such that at least one of the sets $U_{i_0},\ldots,U_{i_m}$ does not intersect V then $U_{i_0} \cap \ldots \cap U_{i_m} \cap (D - V) = U_{i_0} \cap \ldots \cap U_{i_m}$. It is therefore apparent that this covering $\mathcal{U}$ satisfies condition (4); and consequently $H^p(D - V, \mathcal{O}) = 0$ for $1 \leq p \leq d-2$ whenever D is a domain of holomorphy in $\mathbb{C}^n$ and V is a linear subvariety of D with $\dim V = n - d$.

Next consider the special case that D is an open subset of $\mathbb{C}^n$ such that $H^p(D, \mathcal{O}) = 0$ whenever $1 \leq p \leq d-2$ and that V is a complex analytic submanifold of D such that $\dim V \leq n - d$. For any point $a \in D$ choose an open neighborhood U_a of a in D

such that U_a is a domain of holomorphy and such that U_a is sufficiently small that there is a complex analytic homeomorphism $\varphi_a: U_a \to U_a'$ transforming the subvariety $U_a \cap V$ to a linear subvariety of U_a'; any finite intersection of these sets U_a will of course have the same property, and it then follows from the special case considered in the preceding paragraph that the covering $\mathcal{U} = \{U_a\}$ satisfies condition (4). Consequently $H^p(D - V, \mathcal{O}) = 0$ for $1 \leq p \leq d - 2$ whenever D is an open subset of $\mathbb{C}^n$ such that $H^p(D, \mathcal{O}) = 0$ for $1 \leq p \leq d - 2$ and V is a complex analytic submanifold of D such that $\dim V \leq n - d$.

Finally for the general case of the theorem consider an open subset $D \subseteq \mathbb{C}^n$ such that $H^p(D, \mathcal{O}) = 0$ for $1 \leq p \leq d - 2$; it will be proved by induction on the dimension of V that if V is a complex analytic subvariety of D such that $\dim V \leq n - d$ then $H^p(D - V, \mathcal{O}) = 0$ for $1 \leq p \leq d - 2$. If $\dim V = 0$ then V is necessarily a complex analytic submanifold of D, and the desired result follows from the special case considered in the preceding paragraph. If $\dim V > 0$ and it is assumed that the desired result holds for all subvarieties of D having dimension strictly less than the dimension of V, then the desired result holds in particular for the singular locus $\mathcal{S}(V)$ of the subvariety V and consequently $H^p(D - \mathcal{S}(V), \mathcal{O}) = 0$ for $1 \leq p \leq d - 2$; but then $\mathcal{R}(V)$ is a complex analytic submanifold of $D - \mathcal{S}(V)$, and it follows from the special case considered in the preceding paragraph that $H^p(D - V, \mathcal{O}) = H^p((D - \mathcal{S}(V)) - \mathcal{R}(V), \mathcal{O}) = 0$ for $1 \leq p \leq d - 2$. That completes the induction step and concludes the proof of the theorem.

Corollary 1 to Theorem 22. If V is a complex analytic subvariety of an open subset of $\mathbb{C}^n$ and if $\dim V \leq n - d$ where $d \geq 3$ then every point $z \in V$ has arbitrarily small open neighborhoods U such that $H^p(U - U \cap V, \mathcal{O}) = 0$ for $1 \leq p \leq d - 2$.

Proof. If U is any neighborhood of z such that U is a domain of holomorphy in $\mathbb{C}^n$ then $H^p(U, \mathcal{O}) = 0$ for all $p \geq 1$, and the desired result is an immediate consequence of Theorem 22.

The preceding result is all that was required to complete the discussion in §3(d), but a few further remarks will be added here to round out the appendix. If V is a germ of a complex analytic subvariety at the origin in $\mathbb{C}^n$ then the difference $n - \dim V$ will be called the codimension of the germ V and will be denoted by codim V; thus $\dim V + \operatorname{codim} V = n$. Corollary 1 to Theorem 22 can be restated as the assertion that for any sufficiently small open neighborhood U of the origin in $\mathbb{C}^n$ which is also a domain of holomorphy then

$$H^p(U - U \cap V, \mathcal{O}) = 0 \quad \text{for} \quad 1 \leq p \leq \operatorname{codim} V - 2 ;$$

these cohomology groups also vanish for sufficiently large dimensions as well, and the results in this direction can be stated in a conveniently parallel manner in terms of the following definitions. The algebraic codimension of the germ V will be defined as the minimal number of generators of the ideal $\operatorname{id} V \subseteq {}_n\mathcal{O}$, and will be denoted by alg codim V; the geometric codimension of the germ V will be defined as the minimal number r for which there exists an ideal

$\mathfrak{A} \subseteq {}_n\mathcal{O}$ such that $\mathfrak{A}$ is generated by r elements and $\sqrt{\mathfrak{A}}$ = id V, and it will be denoted by geom codim V. Equivalently of course the geometric codimension of the germ V is the minimal number r for which there exist r elements $f_i \in {}_n\mathcal{O}$, all of which can be viewed as holomorphic functions in some open neighborhood U of the origin in $\mathbb{C}^n$, such that the germ V is represented by the analytic subvariety

$$V = \{z \in U | \ f_1(z) = \dots = f_r(z) = 0\} ;$$

or more briefly but less accurately, the geometric codimension of the germ V is the minimal number of functions describing the germ V geometrically. It is clear from the definition that

$$\text{geom codim } V \leq \text{alg codim } V ,$$

and it follows easily from Theorem 9(e) of CAV I that

$$\text{codim } V \leq \text{geom codim } V ;$$

but these inequalities can be strict inequalities. As in §3(f) the germ V will be called an <u>algebraic complete intersection</u> (or just a complete intersection) if codim V = alg codim V; and the germ V will be called a <u>geometric complete intersection</u> if codim V = geom codim V. Any germ which is an algebraic complete intersection must represent a perfect germ of a complex analytic variety, and is also trivially a geometric complete intersection.

Theorem 23. If V is a germ of a complex analytic subvariety at the origin in $\mathbb{C}^n$ then for any sufficiently small open neighborhood U of the origin in $\mathbb{C}^n$ which is also a domain of holomorphy

$$H^p(U - U \cap V, \mathcal{O}) = 0 \qquad \text{for } p \geq \text{geom codim } V .$$

Proof. If $r = \text{geom codim } V$ then for any sufficiently small open neighborhood U of the origin in $\mathbb{C}^n$ there are holomorphic functions $f_1, \ldots, f_r$ in U such that

$$U \cap V = \{z \in U | \; f_1(z) = \ldots = f_r(z) = 0\} .$$

If U is a domain of holomorphy then the sets

$$U_i = \{z \in U | \; f_i(z) \neq 0\}$$

are also domains of holomorphy, and $\mathcal{U} = \{U_1, \ldots, U_r\}$ is a covering of $U - U \cap V$; hence by Lemma 1

$$H^p(U - U \cap V, \mathcal{O}) \cong H^p(\mathcal{U}, \mathcal{O}) \qquad \text{for all } p \geq 0 .$$

However since $\mathcal{U}$ contains only r open sets altogether then for the skew-symmetric cochain groups it follows that $C^p(\mathcal{U}, \mathcal{O}) = 0$ whenever $p + 1 > r$, and consequently $H^p(\mathcal{U}, \mathcal{O}) = 0$ whenever $p \geq r$. That suffices to conclude the proof of the theorem.

Corollary 1 to Theorem 23. If V is a germ of a complex analytic subvariety at the origin in $\mathbb{C}^n$ and if V is a geometric complete intersection then for any sufficiently small open neighborhood U of the origin in $\mathbb{C}^n$ which is also a domain of holomorphy $H^p(U - U \cap V, \mathcal{O}) \neq 0$ only for $p = 0$ or $p = \operatorname{codim} V - 1$.

Proof. It follows from Theorem 22 that

$$H^p(U - U \cap V, \mathcal{O}) = 0 \quad \text{for} \quad 1 \leq p \leq \operatorname{codim} V - 2 ;$$

and since $\operatorname{codim} V = \operatorname{geom\,codim} V$ it follows from Theorem 23 that

$$H^p(U - U \cap V, \mathcal{O}) = 0 \quad \text{for} \quad p \geq \operatorname{codim} V .$$

The only dimensions p for which the cohomology group need not vanish are hence $p = 0$ and $p = \operatorname{codim} V - 1$, and that suffices to conclude the proof of the corollary.

In the situation described in Corollary 1 to Theorem 23 it is obvious that $H^0(U - U \cap V, \mathcal{O}) \neq 0$, and it is quite easy to see that $H^p(U - U \cap V, \mathcal{O}) \neq 0$ for $p = \operatorname{codim} V - 1$ as well. Indeed if that were not the case then $H^p(U - U \cap V, \mathcal{O}) = 0$ for all $p > 0$, and if $\operatorname{codim} V > 1$ that can be shown to lead to a contradiction in the following manner. If W is any other germ of an analytic subvariety at the origin in $\mathbb{C}^n$ and if $\mathcal{J}$ is the sheaf of ideals of W then the coherent analytic sheaf $\mathcal{J}$ has a finite free resolution

$$0 \rightarrow {}_n\mathcal{O}^{r_{n-1}} \rightarrow \cdots \rightarrow {}_n\mathcal{O}^{r_1} \rightarrow {}_n\mathcal{O}^{r} \rightarrow \mathcal{J} \rightarrow 0$$

over some open neighborhood U of the origin; and if U is a domain of holomorphy for which $H^p(U - U \cap V, \mathcal{O}) = 0$ for all

$p > 0$ it is quite clear that $H^p(U - U \cap V, \mathcal{O}) = 0$ for all $p > 0$ as well. Then from the exact cohomology sequence associated to the exact sheaf sequence

$$0 \to \mathcal{J} \to {}_n\mathcal{O} \to {}_W\mathcal{O} \to 0$$

it follows that the restriction mapping

$$\Gamma(U - U \cap V, \mathcal{O}) \longrightarrow \Gamma(W \cap (U - U \cap V), \mathcal{O})$$

is surjective; consequently any holomorphic function on $W \cap (U - U \cap V)$ is the restriction to W of a holomorphic function on $U - U \cap V$. However if $\operatorname{codim} V > 1$ it follows from the extended Riemann removable singularities theorem that any holomorphic function on $U - U \cap V$ extends to a holomorphic function on all of U; and therefore any holomorphic function on $W - W \cap V$ extends to a holomorphic function on all of W, when W is viewed as an analytic subvariety of U. If W is one-dimensional and $W \cap V$ is a point that is obviously not the case though; and it follows from this contradiction that $H^p(U - U \cap V, \mathcal{O}) \neq 0$ whenever $p = \operatorname{codim} V - 1 \geq 1$.

It is not difficult to find examples of germs of complex analytic subvarieties V at the origin in $\mathbb{C}^n$ such that the local analytic cohomology groups of the complement of V are nontrivial in other dimensions than 0 or $\operatorname{codim} V - 1$, hence such that V is not a geometric complete intersection; one approach consists in applying the following general cohomological result to reducible germs of complex analytic varieties.

Lemma 3. (Mayer-Vietoris Sequence) If U is a paracompact Hausdorff space, U_1 and U_2 are open subsets of U such that $U = U_1 \cup U_2$, and $\mathcal{S}$ is a sheaf of abelian groups over U, then there is an exact sequence of groups as follows:

$$0 \longrightarrow H^0(U,\mathcal{S}) \longrightarrow \dots \longrightarrow H^p(U,\mathcal{S}) \longrightarrow H^p(U_1,\mathcal{S}) \oplus H^p(U_2,\mathcal{S}) \longrightarrow$$
$$\longrightarrow H^p(U_1 \cap U_2, \mathcal{S}) \longrightarrow H^{p+1}(U,\mathcal{S}) \longrightarrow \dots \quad .$$

Proof. This is a well known result in various cohomology theories, and the proof for the case of cohomology groups with coefficients in a sheaf is simple enough to be left to the reader; details can be found in the paper by A. Andreotti and H. Grauert, Théorèmes de finitude pour la cohomologie des espaces complexes, Bull. Soc. Math. France, vol. 90, 1962, pp. 193-259 (especially page 236).

To apply this result in the simplest manner, consider two germs V_1, V_2 of complex analytic submanifolds at the origin in $\mathbb{C}^n$ such that the intersection $V_1 \cap V_2$ is also a germ of a complex analytic submanifold; and let V be the reducible germ of a complex analytic subvariety at the origin in $\mathbb{C}^n$ defined by $V = V_1 \cup V_2$. If U is any sufficiently small open neighborhood of the origin in $\mathbb{C}^n$ which is a domain of holomorphy and if $r_i = \operatorname{codim} V_i$ then it follows from Corollary 1 to Theorem 23 that

$$H^p(U - U \cap V_i, \mathcal{O}) \neq 0 \quad \text{only for} \quad p = 0 \quad \text{or} \quad p = r_i - 1 \; ;$$

and if $r = \operatorname{codim} V_1 \cap V_2$ then by the same corollary

$$H^p(U - U \cap V_1 \cap V_2, \mathcal{C}) \neq 0 \quad \text{only for } p = 0 \text{ or } p = r - 1 .$$

Supposing that $r_1 \leq r_2$ it follows that $\dim V = n - r_1$, and hence that $\operatorname{codim} V = r_1$; and then it follows from Corollary 1 to Theorem 22 that

$$H^p(U - U \cap V, \mathcal{O}) = 0 \quad \text{for } 1 \leq p \leq r_1 - 2 .$$

It might be expected that $H^p(U - U \cap V, \mathcal{O}) \neq 0$ for $p = r_1 - 1$; it will be demonstrated that this is the case, and indeed that this cohomology group can be nontrivial for larger indices as well. Of course if $V = V_1$ then $H^p(U - U \cap V, \mathcal{O}) \neq 0$ for $p = r_1 - 1$; excluding this trivial case it can be assumed that $V \subset V_2$, hence that $r > r_2 \geq r_1$. Applying Lemma 3 to the subset

$$U - U \cap V_1 \cap V_2 = (U - U \cap V_1) \cup (U - U \cap V_2)$$

there results an exact cohomology sequence containing the segment

$$H^{r_1-1}(U - U \cap V_1 \cap V_2, \mathcal{O}) \longrightarrow$$

$$\longrightarrow H^{r_1-1}(U - U \cap V_1, \mathcal{O}) \oplus H^{r_1-1}(U - U \cap V_2, \mathcal{O}) \longrightarrow H^{r_1-1}(U - U \cap V, \mathcal{O})$$

The middle term in this segment of exact sequence is nontrivial, while the left hand term is trivial since $r \neq r_1$; and consequently the right hand term must be nontrivial hence

$$H^{r_1-1}(U - U \cap V, \mathcal{O}) \neq 0 .$$

Another segment of the same exact cohomology sequence is

$$H^{r-2}(U-U\cap V,\mathcal{O}) \longrightarrow H^{r-1}(U-U\cap V_1\cap V_2,\mathcal{O}) \longrightarrow$$

$$\longrightarrow H^{r-1}(U-U\cap V_1,\mathcal{O}) \oplus H^{r-1}(U-U\cap V_2,\mathcal{O}) .$$

Again assuming that $V \subset V_2$ so that $r > r_2 \geq r_1$, the right hand term in this segment of exact sequence is trivial; but the middle term is nontrivial, hence the left hand term must be nontrivial as well, or

$$H^{r-2}(U - U \cap V, \mathcal{O}) \neq 0 .$$

If $r-2 > r_1 - 1$ this is indeed an additional nonvanishing cohomology group, and it follows from Theorem 23 that

$$\text{geom codim } V \geq r - 1 > r_1 = \text{codim } V$$

hence that V is not a geometric complete intersection. In general of course $r_2 \leq r \leq r_2 + r_1$, and r certainly can exceed $r_1 + 1$. For instance if V_1, V_2 are two-dimensional submanifolds of $\mathbb{C}^4$ and their intersection is a single point then $r_1 = r_2 = 2$ while $r = r_1 + r_2 = 4$; and in that case geom codim $V \geq 3$ while codim $V = 2$, hence $V = V_1 \cup V_2$ cannot be a geometric complete intersection. An alternative way of seeing this can be found in the paper by R. Hartshorne, Complete intersections and connectedness, Amer. Jour. Math., vol. 84, 1962, pp. 497-508.

INDEX OF SYMBOLS

	Page
ann S	117
ass A	118
$\text{hom dim}_{V^{\mathcal{O}}} A = \text{hom dim}_V A$	105
$\text{hom dim}_{\varphi} V$	110
hom dim V	110
$\text{hom dim } V_p$	116
$\text{hom dim}_k A$	133
$\text{prof}_{V^{\mathcal{O}}} A = \text{prof}_V A$	121
prof V	127
$\text{prof}_k A$	133
syz A	102
$\text{syz}^n A$	104

(Also see page 164 of Lectures on Complex Analytic Varieties: The Local Parametrization Theorem.)

INDEX

A-sequence, 120

annihilator of a subset of an ${}_V\mathcal{O}$-module, 117

associated prime ideal of an ${}_V\mathcal{O}$-module, 118

base point of a germ of complex analytic variety, 3

branched analytic covering, generalized, 12

branching order of a finite analytic mapping, 25

characteristic ideal of an analytic mapping, 19

codimension, homological, 131

complete intersection, 140, 142

-----, algebraic, 153

-----, geometric, 153

conductor of a germ of complex analytic variety, 32

covering, generalized branched analytic, 12

denominator, universal, 28

dimension, homological, 105, 110

direct image sheaf, 17

divisors, zero, 118

equivalent germs of complex analytic subvarieties, 2

equivalent germs of complex analytic varieties, 3

finite analytic mapping, 11

finite analytic mapping of branching order r, 25

finite homomorphism, 16

generalized branched analytic covering, 12

germ of complex analytic subvariety, 1

germ of complex analytic variety, 3

germ of holomorphic function, 5

homological codimension, 131

homological dimension of an ${}_V\mathcal{O}$-module, 105

homological dimension of a germ of complex analytic variety, 110

homological resolution of an ${}_V\mathcal{O}$-module, 104

ideal, associated prime, 118

ideal, characteristic, 19

intersection, complete, 140, 142

length of an A-sequence, 120

local ring of a germ of complex analytic variety, 5

maximal A-sequence,

minimal free (homological) resolution of an ${}_V\mathcal{O}$-module, 104

normal germ of a complex analytic variety, 34

normalization of a germ of complex analytic variety, 34

order, branching, 25

order of a holomorphic function along a submanifold, 51

order of a holomorphic function along a subvariety, 53

order of a meromorphic function along a submanifold, 52

perfect germ of a complex analytic variety, 94

point, base, 3

prime ideal, associated, 118

profundity of an ${}_V\mathcal{O}$-module, 121

profundity of a germ of complex analytic variety, 127

removable singularity set for holomorphic functions, 96

resolution, minimal free (homological), 104

ring, local, 5

ring of germs of holomorphic functions, 5

sheaf, direct image, 17

simple analytic mapping, 43

singularity, removable, 96

subvariety, complex analytic, 1

-----, equivalent germs of, 2

-----, topologically equivalent germs of, 3

syzygy module of an ${}_V$ -module, 100

universal denominator, 28

variety, complex analytic, 6

-----, germ, 3

zero-divisor for an ${}_V\mathcal{O}$-module, 118

Library of Congress Cataloging in Publication Data

Gunning, Robert Clifford, 1931-
Lectures on complex analytic varieties: finite analytic mappings.

(Mathematical notes, 14)
1. Analytic mappings. 2. Analytic spaces.
I. Title. II. Series: Mathematical notes (Princeton, N. J.), 14.
QA331.G783 1974 515'.9 74-2969
ISBN 0-691-08150-6